10 660 8085

Perfumes, Cosmetics and Soaps

VOLUME II The Production, Manufacture and Application of Perfumes

W. A. POUCHER

Ninth edition

CHAPMAN & HALL
London · Glasgow · New York · Tokyo · Melbourne · Madras

Published by Chapman & Hall, 2-6 Boundary Row, London SE1 8HN

Chapman & Hall, 2-6 Boundary Row, London SE1 8HN, UK

Blackie Academic & Professional, Wester Cleddens Road, Bishopbriggs, Glasgow G64 2NZ, UK

Chapman & Hall, 29 West 35th Street, New York NY10001, USA

Chapman & Hall Japan, Thomson Publishing Japan, Hirakawacho Nemoto Building, 6F, 1-7-11 Hirakawa-cho, Chiyoda-ku, Tokyo 102, Japan

Chapman & Hall Australia, Thomas Nelson Australia, 102 Dodds Street, South Melbourne, Victoria 3205, Australia

Chapman & Hall India, R. Seshadri, 32 Second Main Road, CIT East, Madras 600 035, India

---

First edition 1923
Ninth edition 1993

© 1993 Chapman and Hall

Printed in Great Britain by TJ Press, Padstow, Cornwall

ISBN 0 412 27350 0

Apart from any fair dealing for the purposes of research or private study, or criticism or review, as permitted under the UK Copyright Designs and Patents Act, 1988, this publication may not be reproduced, stored, or transmitted, in any form or by any means, without the prior permission in writing of the publishers, or in the case of reprographic reproduction only in accordance with the terms of the licences issued by the Copyright Licensing Agency in the UK, or in accordance with the terms of licences issued by the appropriate Reproduction Rights Organization outside the UK. Enquiries concerning reproduction outside the terms stated here should be sent to the publishers at the London address printed on this page.
  The publisher makes no representation, express or implied, with regard to the accuracy of the information contained in this book and cannot accept any legal responsibility or liability for any errors or omissions that may be made.

A catalogue record for this book is available from the British Library

UNIFORM WITH THIS VOLUME

W. A. Poucher's
Perfumes, Cosmetics and Soaps

Volume I   The Raw Materials of Perfumery
Volume III Cosmetics

Preface

During the past decade there have been many changes in the perfumery industry which are not so much due to the discovery and application of new raw materials, but rather to the astronomic increase in the cost of labour required to produce them.

This is reflected more particularly in the flower industry, where the cost of collecting the blossoms delivered to the factories has gone up year after year, so much so that most flowers with the possible exception of Mimosa, have reached a cost price which has compelled the perfumer to either reduce his purchases of absolutes and concretes, or alternatively to substitute them from a cheaper source, or even to discontinue their use.

This development raises an important and almost insoluble problem for the perfumer, who is faced with the necessity of trying to keep unchanged the bouquet of his fragrances, and moreover, to ensure no loss of strength and diffusiveness. Of course, this problem applies more especially to the adjustment of formulae for established perfumes, because in every new creation the present high cost of raw materials receives imperative consideration before the formula is approved.

The makers of artificial flower oils have for years anticipated this change and have directed their research to the discovery and synthesis of materials which will improve their own flower compounds and bring them nearer to the duplication of the characteristic bouquets of the naturals. And several of them have succeeded so well that the replacement of part, or even of the whole, of the naturals in finished perfumes has been facilitated. Whether these gradual alterations have been noticed by the users may be problematical, but the alternative would have been to maintain the original formulae and just increase the sale price.

Another trend which started many years ago with the smaller perfumery houses has spread throughout the industry and is now common practice with almost all firms, save a few of the distinguished Paris perfumers. It is well known that the more important employed their own perfumers who created the fragrances they marketed, whereas today it is not uncommon for them to buy finished compounds. The usual procedure is to ask for samples from the firms specialising in this work and to choose one of them which may be slightly modified or merely diluted with alcohol and after maturing placed on the market. This not only eliminates the high cost of research but also the stocking of a vast array of expensive raw materials. It has resulted in great competition to secure the services of gifted perfumers, most of whom now work for one or other of the houses creating these compounds.

Throughout its many editions this work has been invaluable to perfumery research and should continue to be useful even in the changed conditions adumbrated above.

W. A. Poucher

4, Heathfield,
Reigate Heath, Surrey

Contents

		page
Preface		vii
1	Historical Sketch	1
2	The Production of Natural Perfumes	16
3	The Purchase and Use of Flower Absolutes	41
4	Odour Classification and Fixation	46
5	Monographs on Flower Perfumes	68
6	Miscellaneous Fancy Perfumes	217
7	Toilet Waters	246
8	Soap Perfumery	268
9	Tobacco Flavours	321
10	Floral Cachous	332
11	Incense and Fumigants	339
12	Sachets and Solid Perfumes	344
13	Fruit Flavours	351
Appendix		361
Index		373

CHAPTER ONE

Historical Sketch

If it were possible to delve into the past at a sufficiently remote period, it would probably be found that the romance of perfumery had its beginning with the Atlantians who flourished at a period conjectured to antedate the Christian Era by about 23,000 years. Cosmetics appear to have been known to this second sub-race, who are believed to have employed a form of petroleum as one of their principal toilet accessories.

The Chinese may have been the forerunners of Western civilisation, and although they are believed to have conquered the aboriginal tribes who inhabited that part of Asia some time during the third Millennium B.C., little is known concerning their history before 800 B.C.

It is therefore necessary to turn to Egypt for the earliest records of perfumery. The first Dynasty is known as the Thinite and its first ruler was King Menes. He is said to have conquered Lower Egypt, founded Memphis, and built the temple of Ptah. His tomb was opened in 1897. Other tombs of the eight kings of the first Dynasty and of the nine kings of the second are at Abydos, and all show traces of the Egyptian habit of burying needments and luxuries for the dead. The first Dynasty is variously placed at 3500 to 5000 years B.C. and several examples of art at that period still exist. For instance, in the British Museum, there are many beautiful unguent vases carved in alabaster which authorities have dated about 3500 B.C. Other specimens of interest to the perfumer are:

Mirrors used in the sixth Dynasty—2800 B.C.

Kohl vases (in glass) and stibium pencils used in the eighteenth Dynasty—1500 B.C.

Papyrus showing men and women having lumps of nard fixed on top of the head—1500 B.C.

The opening by Mr Howard Carter of the tomb of Tutankhamen who ruled about 1350 B.C. has brought to light many excellent specimens of the early perfumer's art. According to eyewitnesses, the unguent vases, exquisitely executed in alabaster, contained quantities of aromatics which were still elusively fragrant. This cosmetic was examined in 1926 by Chapman and Plenderleith. The odour emitted has been compared variously with cocoa-nut oil, broom, and valerian. The chemical evidence supported the view that the fat was of animal character and accounted for 90 per cent of the whole. The remaining 10 per cent appeared to consist of some resin or balsam.

Cosmetics, manicure instruments, and razors thousands of years old are stated to be among the latest objects forwarded from the tomb of Queen Hetepheres to the Cairo Museum. Hetepheres was the mother of Khufu, or Cheops, the Pharaoh of the fourth Dynasty about (3500 B.C.?) to whom is attributed the Great Pyramid at Giza.

The Egyptian Office of Works communique says that the articles include thirty alabaster vessels, a large copper ewer with its copper basin and toilet box, three gold cups and implements and tools of gold, copper, and flint. Among the alabaster vessels are two of unique form. The toilet box of wood is a reconstruction of an old box which was found in fragments on the floor, but the contents are the original contents, eight small alabaster jars and a copper spoon. Seven of the jars contained the seven traditional perfumed ointments of the Egyptians and the eighth contained kohl. Six of the lids of these jars have been preserved and inscribed with the names of the contents, while a single hieroglyphic sign on the rim of each jar indicates the connection between each lid and its respective jar. The contents of the jars consist of dry fibrous remains, probably vegetable, which have been removed for examination and analysis.

Objects in solid gold include a small drinking cup with a re-curved rim and spout, two small cups, two razors, three rectangular knives, a manicure implement with a sharp end for cleaning the nails and a rounded end for pressing down the skin at the base of the nails. The copper implements consist of five razors, which, with the two gold razors, make a set of seven, and four rectangular knives which, with the three gold knives, make another set of seven. With these is a set of extraordinary flint implements, which seem to be older prototypes of the metal implements, thirteen oval flints or flint razors, and nine rectangular flint knives. There is also a very fine, small copper needle.

On other monuments and tombs in Egypt there is still ample evidence of their great esteem for aromatics. For instance, on the large granite tablet inserted in the breast of the sphinx, King Thothmes IV (about 1600 B.C.) is portrayed making an offering of incense and of fragrant oil or unguent.

Perfumes were used for three quite distinct purposes by the Egyptians:

1. As offerings to their deities.
2. For aesthetic purposes during their lives.
3. As the principal agents for embalming their dead.

It was customary for the priests to burn incense before the gods in the temples, and this incense probably consisted of aromatic gums, resins, and oleo-resins mixed with perfumed woods. These substances were made into small pieces and were volatilised by being thrown into the glowing censers. In some cases these operations were performed several times a day. At Heliopolis— where the sun worshippers foregathered—resins were burned at dawn, myrrh at noon, and kaphi at sunset. This substance *kaphi* is believed to have been a mixture of several aromatic ingredients, of the nature of which there is no record. At the féte of the God, Isis, it was customary to sacrifice an ox. The odour of burnt flesh was, however, so obnoxious, that the worshippers found it necessary to fill up his interior with aromatic gums and oils, which made these sacred observances more amenable. In religious processions there was always a lavish display of perfumes, while no king was ever crowned without being anointed with fragrant oils by the priests. At this period it is probable that the priests made most of the perfumes, fragrant oils, and unguents; they were therefore the perfumers of their time and their pursuit was considered a mysterious and much-esteemed art. The containers were beautiful *objets d'Art* and were executed in all kinds of valuable material. Ivory and alabaster were the principal substances used, while frequently carved wood, onyx, and porphyry were fashioned into pots and vases.

The ancient Egyptians were probably the inventors of the bath, which habit, in later years, was treated on a much more elaborate scale by the Greeks and Romans. This form of ablution was probably necessitated by the terrific heat of that country and it was followed by the liberal application of perfumed oils and unguents. Doubtless these were employed to give the skin more elasticity as well as to impart a balmy and pleasing effect to these aesthetic people. This practice, though not so extensive, is still applied by the Eastern nations in order to prevent an undue drying of the epidermis and the irritation that would follow.

Sesame oil appears to have been one of the most favoured vehicles for the aromatics, although both almond and olive oils were undoubtedly used. Some of the odorous constituents were grown in Egypt while the greater proportion were most probably imported from Arabia. Amongst the former may be mentioned thyme and origanum, together with a substance called *balanos*, which appears to have been extracted from the shells of some unidentified fruit, while of the latter myrrh, olibanum, and spikenard were of great importance.

The process of embalming the dead was so important during the Egyptian era that sections of the towns were peopled by men who made this their profession. According to Herodotus, who travelled much in Egypt, it was customary to remove the brains and intestines, replacing the former with drugs and the latter with aromatic gums such as myrrh and frankincense. The body was then covered with native sodium sesquicarbonate (natron) for about two months (seventy days) which dried up the epidermis and the underlying tissues. Linen was soaked with gums and the body wrapped up in this prepared material. The whole was then enclosed in a timber case fashioned somewhat in the human form and subsequently painted in bright colours. This was the process adopted by the élite of Egyptian society and was by no means an inexpensive matter. The common people who could not afford this process had their dead preserved by the injection of oil or by merely the soda bath; the corpse being subsequently enclosed as above indicated.

The use of cosmetics had not escaped the attention of the Egyptian ladies, who enhanced their personal beauty by the employment of somewhat crude paints. These practices reached their zenith in the time of Cleopatra. It seems probable, from discoveries in Egyptian tombs, that the highest degree of cosmetic art was attained in the embellishment of their eyes. This effect was produced by painting the under side of the eye green and the lid, lashes and eyebrows black by the application of *kohl*—the product being made chiefly from galena and applied with an ivory or wooden stick. Imitations of these prettily worked implements, together with the kohl boxes, are sold today in Egypt. Combs and mirrors were also used by the Egyptian ladies. According to A. Lucas,[1] out of 61 samples of ancient Egyptian kohl analysed, 40 contained approximately 65 per cent of galena, a lead ore. Two of these contained a trace of antimony sulphide and 4 some carbon ... of the remaining 21 samples, 2 consisted of lead carbonate, 1 black oxide of copper, 5 brown ochre, 1 magnetic oxide of iron, 6 manganese oxide, 1 antimony sulphide, 4 malachite, a copper ore, and 1 chrysocolla, a greenish-blue copper ore.

Lip-salves, probably dating back to about 3500 B.C., have been brought to light by Dr A. Kenneth Graham during excavations at Ur. This gentleman is of the opinion that they were used by Queen Shubad, and were found to contain large quantities of lead.

The use of henna was (and still is) much in favour and was applied to the finger-nails and palms of the hands. Good and well-preserved specimens can be seen today in the British Museum.

At the banquets and entertainments given by the wealthy Egyptians, it was customary to have a lavish display of flowers. This consisted in strewing the rooms and tables with lotus and saffron flowers. A peculiar custom, related by Herodotus, consisted in the introduction to these banquets of a mummy. A man would enter when the revel was at its height and shout, 'Look at this. Drink and make merry, for so you will be after your death.'

The first chapters of the Bible are believed to refer to about 4004 B.C., and in the description of the Garden of Eden[2] there is a reference to *bdellium*. It has been impossible to determine the constitution of this substance from this particular mention, but it has been presumed to be even a mineral by reason of its association with onyx stone. It was about 2120 B.C. that Abram journeyed into Egypt,[3] and as there is only the above-mentioned

---

[1] *Ancient Egyptian Materials and Industries.*   [2] Gen. ii. 12.   [3] Gen. xii. 14.

reference to aromatic substances before this date, it must be presumed that the Hebrews obtained their knowledge of the uses of perfumes from the Egyptians. The earliest specific reference to commerce in aromatic substances is about 1730 B.C. when the Ishmeelites came from Gilead with their camels bearing spicery and balm and myrrh.[4] Exactly what is meant by spicery has not been definitely proved, but according to different commentators, it may be considered to be either storax or tragacanth. The constitution and source of *balm* is equally indefinite. It is considered by some botanists to be mastic because this plant is prolific in the rocky country of Gilead. Others are of the opinion that it was derived from a species of the N.O. Simarubaceae. By pounding and boiling the fruit of this plant an oil was obtained which was known as 'Balm of Gilead'.

In the Book of Exodus[5] the references to aromatic substances become specific and are numerous.[6] The Shittim wood mentioned in the first verse was probably obtained from a species of *Acacia*. The spices referred to in verses 23, 24, and 25 (containing the instructions for the preparation of holy ointment) are not grown in Egypt, and it must be assumed therefore that they were brought from the east by the traders. The *stacte* and *onycha* mentioned in verse 34 are supposed to have been styrax and labdanum respectively. Other references to aromatic substances are to be found in Exodus xxxv, xxxvii, xl, Numbers xvi, 2 Kings ix, 2 Chronicles ix, xxvi, Esther ii, Psalms xlv, cxxxiii, Proverbs vii, xxvii, Isaiah xxxix, lvii, St. Mark xiv, St. John xix. The Song of Solomon is dated about 1014 B.C. and it contains numerous references to perfumes. Most of these are mentioned elsewhere in the Scriptures, with the exception of *camphire*.[7] This is considered by botanists to be henna, or at least a closely allied species of *Lawsonia*, possibly *L. alba*.

Cosmetics were evidently used by the Jewish women, for it is written that 'when Jehu was come to Jezreel, Jezebel heard of it; and she painted her face, and tired her head, and looked out at a window'.[8] This is explained more clearly in the following passage, 'Thou didst wash thyself, paintedst thy eyes,[9] and deckedst thyself with ornaments'.[10]

---

[4] Gen. xxxvii. 25.   [5] Chapter xxx.   [6] Verses 1, 7, 23, 24, 25, 34, 35, 36, 37, 38.
[7] Song of Solomon i. 14, iv. 13.   [8] 2 Kings ix. 30.   [9] Probably with kohl.
[10] Ezek. xxiii. 40, and see also Jer. iv. 30.

HISTORICAL SKETCH

While briefly referring to perfumery in connection with religion and the Scriptures, it seems desirable that something should be said about the Koran, although it was written several centuries later—about A.D. 600. Mahomet wrote this work in a very disconnected manner, and the chapters are called Suras. In these Suras appear frequent references to aromatic substances, and there is no doubt that the Arabs' love of perfumes helped them to appreciate the teachings of this religious work to a much greater degree. Musk was, of course, one of their most esteemed perfumes. In Sura lxxxiii[1] it is mentioned as follows:

'The Seal of musk. For this let those pant who pant for bliss.'

In Sura lvi Mahomet describes some of the joys which the faithful shall experience in the Gardens of delight. He says:

'Of a *rare* creation have we created the Houris, and we have made them even virgins.'

In Rodwell's translation the word 'rare' is in italics, but no explanation is given of its meaning.

In Sura lv 'these damsels with retiring glances' are described as having eyes like 'hyacinths and pearls'.

In Suras lii and lvi appear the following passages:

'On couches ranged in rows shall they recline; and to the damsels with large dark eyes will we wed them.'

'And theirs shall be the Houris, with large dark eyes, like pearls hidden in their shells.'

These two passages would probably indicate the application of kohl which was, and still is, so much used by the ladies of the East.

The founder of one of the earliest Persian religions was Zoroaster, who flourished about 1000 B.C. Zoroastrianism is today represented in India by the Parsees. This prophet substituted fire worship for idolatry. Incense was burnt continuously, and to ensure against the fire burning out the priests were changed five times daily.

These ancient Asiatic nations had a great predilection for cosmetics, and this was by no means confined to the fair sex. It is probable that Babylon and Nineveh (corresponding with the modern Paris and London?) were the chief centres where this art was practiced. Herodotus says that the Babylonians were great consumers of aromatics and perfumed their bodies with the most

[1] Rodwell's translation (J. M. Dent & Sons, Ltd.).

expensive odours. A peculiar practice then in vogue was the use of pumice-stone, rubbed on the skin to make it smooth. Red and white paint for the face was common, the former being vermilion and the latter white lead. Egyptian kohl was also a favourite, named at that time stibium, which was probably a sulphide of antimony. This, however, according to A. Lucas is a mistaken idea, possibly arising from the Roman use in eye cosmetics and eye medicines of an antimony compound called by Pliny *stimmi* and *stibi*. It was applied to the lids and corners of the eyes to make them more brilliantly lustrous.

The Greeks held the use of perfumes in high esteem, and the art was practised largely by women. As was usual with all their pursuits, they had some mythological conception concerning the origin of perfumes. Venus was believed to have been the first user of aromatics, and man's knowledge of them was attributed to an indiscretion of one of her nymphs by the name of Aenone. Paris thus conveyed to Helen of Troy the secret which enabled her to acquire and retain that marvellous beauty for which she was so famed. Homer frequently refers to perfumes in his 'Iliad' and 'Odyssey'. In the former he thus describes the toilet of Juno:

> Here first she bathes, and round her body pours
> Soft oils of fragrance and ambrosial showers.

Theophrastus was probably the earliest Greek writer on the subject of perfumery. He was born in 370 B.C. and lived to the age of eighty-five. His principal work was on botany and is characterised by a peculiar and yet remarkable classification of plants. His minor works on perfumery and the weather are equally interesting reading. For instance, he speaks of a compounded perfume (as distinct from a flower perfume) as one that is artificially and deliberately produced; thus the method of the makers of perfumed powders is to mix solid with solid, that of those who compound unguents is to mix liquid with liquid: but the third method which is commonest, is that of the perfumer, who mixes solid with liquid.

Concerning fixation, Theophrastus says:[12] 'Now the composition and preparation of perfumes aim entirely, one may say, at making odours last. That is why men make oil the vehicle of them, since it keeps a very long time and also is most convenient for use.'

'They use spices in the making of all perfumes; some to

---

[12] Hort's translation in the Loeb classical library. William Heinemann.

thicken[13] the oil, some in order to impart their odour. The less powerful spices are used for the thickening, and then at a later stage they put in the one whose odour they wish to secure. For that which is put in last always dominates, even if it is in small quantity; thus if a pound of myrrh is put into a half-pint of oil, and at a later stage a third of an ounce of cinnamon is added, this small amount dominates.'

Theophrastus thus describes the raw materials from which perfumes were prepared:

'Perfumes are compounded from various parts of the plants: flowers, leaves, twigs, root, wood, fruit, and gum; and in most cases the perfume is made from a mixture of several parts. Rose and gilliflower perfumes are made from the flowers: so also is the perfume called *Susinon*, this too being made from flowers, namely lilies: also the perfumes named from bergamot mint and tufted thyme, *kypros*, and also the saffron perfume. The crocus which produces this is best in Aegina and Cilicia. Instances of those made from the leaves are the perfumes culled from myrtle and dropwort: this grows in Cyprus on the hills and is very fragrant: that which grows in Hellas yields no perfume being scentless.

'From roots are made the perfumes named from iris, spikenard, and sweet marjoram, an ingredient in which is *koston*;[14] for it is the root to which this perfume is applied. The Eretrian unguent is made from the root of *kypeiron*, which is obtained from Cyclades as well as from Enboea. From wood is made what is called "palm perfume"; for they put in what is called the "spathe", having first dried it. From fruits are made the quince perfume, the myrtle and the bay. The "Egyptian" is made from several ingredients, including cinnamon and myrrh.'

A compound perfume mentioned by this ancient author is one called *Megaleion*, named presumably after its inventor. It contains burnt resin, oil of *balanos*, cassia, cinnamon, and myrrh. This is alleged to have been a very difficult perfume to make!

Concerning the habits of the Greek perfumers Theophrastus says: 'Perfumers seek upper rooms which do not face the sun, but are shaded as much as possible. For the sun or a hot place deprives the perfumes of their odour, and in general makes them lose their character more than cold treatment.'

---

[13] This probably means to reduce its volatility and so make it more persistent.
[14] Probably Kostus.

Other classical literature of the Greeks gives a fairly comprehensive insight into the use they made of cosmetics. In addition to the above-mentioned articles, they coloured their cheeks and lips with a root called *polderos*, which was probably similar to our present alkanet. Hair dye was also known at that period.

The Romans, during their early history, showed very little interest in perfumes, and when King Antiochus and Asia were subdued, an edict was published in Rome (about 188 B.C.) forbidding anyone to sell exotics (unguents). It was only after their migrations into southern Italy, then occupied by the Greeks, that they acquired a more intimate knowledge of the aesthetic side of life. Nero became Emperor of Rome in A.D. 54, and by this time both perfumes and cosmetics had assumed an important rôle at his court. He personally used cosmetics liberally, and his wife Poppaea made no secret of the artifices of the toilet. Amongst the many things they used were white lead and chalk to whiten the skin; Egyptian kohl for the eyelids and lashes; *fucus*, a sort of rouge, for the cheeks and lips; *psilotrum*, a kind of depilatory; barley flour and butter as a cure for pimples, and pumice-stone for whitening the teeth. The ultra-fashionable ladies of the Roman court devised a method for bleaching their hair by means of a sort of soap which came from Gaul. The centre of the perfume and unguent industry in Italy was situated at Capua.

According to Suetonius[15] at the funeral of Poppaea, Nero consumed more incense than Arabia could produce in ten years.[16] In his golden palace the dining-rooms were lined with movable ivory plates, concealing silver pipes, which sprayed on the guests a stream of highly odorous perfumes. It is, of course, well known that the Romans made considerable use of the bath and erected some of the finest bathing establishments, ruins of which can still be seen.

The Romans devised all sorts of beautiful containers for their perfumes and unguents, of which there were three principal kinds:

1. Solid unguents, or *hedysmata*.
2. Liquid unguents, or *stymmata*.
3. Powder perfumes, or *diapasmata*.

The solid unguents were generally of one specific perfume, such as almond, rose, or quince. The liquid unguents were most

[15] Book VI. [16] Pliny says one year, Book XII, Chap. 41.

frequently compounds containing flowers, spices, and gums, and followed very much on the lines quoted above from Theophrastus. The constituents were generally digested in one or other of the fixed oils, such as sesame, olive, or ben.[17] According to Pliny,[18] resin and gum were added to fix the odour in solid perfumes; 'indeed', he says, 'it is apt to die away and disappear with the greatest rapidity if these substances are not employed'. This writer also notes that unguents improved with age and for that reason were stowed away in lead containers. They were tested on the *back* of the hand and not on the palm, owing to the heat thereof having a bad effect on them.

In the centuries that followed the Arabs seem to have delved into the serious side of perfumery more than any other race. For instance, in the tenth century an Arabian doctor by the name of Avicenna made efforts to extract the perfume from flowers by distillation, which at that time was imperfectly understood. He was fortunate enough, however, to isolate from the rose some of its perfume in the form of oil or otto (attar) and to produce also supplies of Rose Water. Quite a trade in this latter article was developed in later years by the Arabs.

The Oriental women lead a somewhat secluded life, and one of their principal interests is in the enhancement of their personal beauty. As a toilet article of importance kohl probably stands pre-eminent. Of secondary importance is a complexion powder named *batikha* which is made from powdered marble, rice, borax, etc.

Concerning the history of perfumes and cosmetics in Britain there is no doubt but that, in a crude way, the early inhabitants of this country derived some pleasure from adorning their body with paints, etc. The Picts undoubtedly obtained their name from *picti*—painted, because of the elaborate designs they painted on their persons.

The importation of perfumes from the East dates from the time of the crusades, when the knights, returning from their conquest, brought with them many of the much-prized toilet articles used by the ladies of the harem. From this time until the reign of Queen Elizabeth perfumes and cosmetics became more and more popular;

---

[17] One of the earliest Roman nobles invented Frangipanni, and this is the only perfume which has withstood the test of time. A monograph on it appears in Vol. I.

[18] Book XIII, Chap. 2.

this vogue was not confined to our island but spread equally quickly throughout France and Italy. A peculiar habit, which has been attributed to Frangipanni, was the wearing of perfumed gloves—the trades of perfumer and glover having become one. About the same period the first alcoholic perfume appeared and is still known as Hungary Water.[19] The origin of this is attributed to Queen Elizabeth of Hungary who first prepared it in the year 1370. At the court of Queen Elizabeth of England both sexes made use of every kind of aromatic for perfuming the body and many new kinds of cosmetics for adorning the person. Powder and patches were all the rage and will always be associated with the reign of this extravagant queen. The toilet preparations employed by the ladies of the court were kept in strongly perfumed boxes called Sweet Coffers. These were considered a necessary part of the furniture of their bedrooms. A recipe for making a beautiful complexion at this time was to first take a very hot bath to induce excessive perspiration, and this was followed by washing the face with plenty of wine to make it fair and ruddy. This latter treatment was even in those days fairly expensive, but Mary Queen of Scots is alleged to have even bathed in wine on which account she applied for an increased allowance! This luxurious habit was not uncommon with the elder ladies of the court, but the younger ones had apparently to be content with milk.

With the advent of the Commonwealth all these luxuries were discarded, but as soon as Charles II was restored to the throne they became popular again. In later years the ladies of the court adopted a new practice by popularising powdered hair, but this soon fell into disuse. By the seventeenth century cosmetics were used to such an extent by nearly all classes that in 1770 an Act was introduced into the English Parliament which was intended to afford some protection to those men who were beguiled into matrimony by the artificial adornments of the fair sex. This Act is as follows:

'That all women, of whatever age, rank, profession, or degree, whether virgins, maids, or widows, that shall, from and after such Act, impose upon, seduce, and betray into matrimony, any of his Majesty's subjects, by the scents, paints, cosmetic washes, artificial teeth, false hair, Spanish wool,[20] iron stays, hoops, high-heeled

---

[19] Consult the chapter on Toilet Waters for manufacturing details.
[20] Consult Carthamin in Vol. I.

shoes, bolstered hips, shall incur the penalty of the law in force against witchcraft and like misdemeanours and that the marriage, upon conviction, shall stand null and void.'

The town of Grasse soon sprung into fame as the seat of the natural flower perfume industry, and by reason of its ideal situation has been able to maintain its premier position. Attempts have been made in other parts of the world to cultivate different flowers for the extraction of their perfume, but beyond the production of experimental quantities, there is, as yet, no serious competitor. In the course of time numerous fragrant oils have been distilled from plants in other parts of the world, and, together with the more recently introduced synthetics and natural isolates, constitute the very wide range of raw materials from which the perfumer blends the alluring odours so delightfully marketed in the many exquisitely designed containers and packages.

Today the use of cosmetics is almost universal, owing no doubt to the greater artistry with which they are employed and also in no less degree to the greater skill and knowledge with which they are prepared. The use by the manufacturer of dangerous ingredients has been discontinued, and cosmetics can therefore be said to constitute a very valuable toilet asset.

To review briefly their use:

first—they are intended to cleanse;
second—to allay skin troubles;
third—to cover up imperfections;
and lastly—to beautify.

To take a simple example of each type will show they all have a legitimate use. Cold cream consists of a finely divided emulsion of fat or oil, which, when applied to the epidermis, removes all adherent dust and skin debris; when it is subsequently rubbed off, the skin is left fresh and white—incidentally healthier for the application. It functions in another way also, for when applied after some kinds of soap, which contain free alkali, it neutralises any drying effect the soap may have had and keeps the skin in a supple condition. Other products having a similar action are creams containing fairly large proportions of lanolin. This fat is closely allied to the natural fat of the skin, so that it acts as a food to the skins of those persons who have a deficiency of natural fat.

In such subjects, eczema is prone to develop—thus lanolin creams undoubtedly help to prevent skin troubles.

Then again, there are talcum powders, which are intended primarily to allay any irritation of the skin. This trouble becomes much more apparent in hot climates, where the consumption of this product is tremendous. In quite another direction there are mouth washes and dental creams. They contain such substances as thymol, which cleanse by reason of their action on the micro-organisms present at all times in the mouth.

It is in connection with the third and fourth uses, however, that the alleged harmfulness of cosmetics finds expression. The arguments generally advanced may be reduced to three—first, that by the excessive use of inferior face creams and powders the pores become enlarged and the skin flabby in consequence; second, that undue friction is necessary in the use of rouge; and third, that cosmetics contain ingredients which are injurious.

To take the last argument first: if these products consisted of such substances as white lead, mentioned earlier in this chapter, there would be some justification for the assertion, but they do not. Some so-called 'skin-beautifiers' contain small quantities of corrosive sublimate which in itself is poisonous, but the minute quantities employed act rather as a skin antiseptic.

Concerning the second: ladies who want to use rouge will use it, no matter what advice may be given to the contrary. It has a tendency to block the pores, and will do no harm if removed within a few hours. Lipsticks as now made have their melting-point such that they soften immediately on application to the lips—hence the amount of friction is so small that it can be disregarded. At one time it was customary for ladies to rely on complexion pills for the 'bloom of youth', but today they find it much more satisfactory to spend their money on external applications which do give the desired result, rather than on internal remedies from which results are frequently problematical.

Perhaps it is in connection with the first argument that most misconception exists. Face creams and, in particular, vanishing creams, are much used as a basis for face powder. They give to the skin a matt appearance and so cover slight imperfections. It is just possible that some of them are of such a consistency that they do block the pores temporarily, but the best are so compounded that they act as cleansers as well as powder creams. The film covering is so thin that it cannot possibly do harm, but on the contrary acts as a protecting agent against the sun. When face powder is lightly dusted on, the evaporation surface is increased, and a consequent cooling during the evaporation of perspiration is experienced.

Like all other things, the use of cosmetics carried to excess might be harmful, but the fact must not be denied that those who take the trouble to enhance their personal attractiveness are a much greater asset to this world than those who are slovenly and neglect their bodies.

Face massage is much appreciated by those who desire to retain a clean, supple, and unwrinkled complexion, but modern surgical science has evolved a treatment known as face lifting, by which wrinkles can be removed or filled up with plastic paraffin base, the shape of the nose improved, and defects to the chin and ears eliminated. The operation is performed with local anaesthetics under highly antiseptic conditions and without danger.[21] This method of rejuvenescence does not, however, pretend to dispense with the use of cosmetics, since plastic surgery only improves contour. A lady of fifty years can lose by this means twenty years of her apparent age, but even so she will use cosmetics to enhance her tightened skin and to give her complexion just that *éclat* which is demanded by the clothes she wears and the society she cultivates.

---

[21] In view of the number of quacks who practice, it is best to entrust one's face to the hands only of the qualified plastic surgeon.

CHAPTER TWO

The Production of Natural Perfumes

The perfume in the plant

Natural perfumes, one of the most marvellous phenomena of plant metabolism, probably reach their highest degree of excellence in the fragrance exhaled by fresh flowers. This fragrance is due to the minute traces of essential oil which exist in the petals, sometimes in the free state, as in rose and lavender, and occasionally in the form of a glucoside which, under favourable conditions, is decomposed in the presence of an enzyme or ferment, as in jasmin and tuberose. The existence of a volatile oil, however, is by no means confined to the inflorescence, but frequently occurs in other parts of the vegetable organism.

For example, it is found in the

Flowers of, cassie, carnation, clove, hyacinth, heliotrope, mimosa, jasmin, jonquille, orange blossom, rose, reseda, violet, and ylang-ylang.

Flowers and leaves of, lavender, rosemary, peppermint, and violet.

Leaves and stems of, geranium, patchouli, petitgrain, verbena, and cinnamon.

Barks of, canella, cinnamon, and cassia.

Woods of, ceda, linaloe, and santal.

Roots of, angelica, sassafras, vetivert.

Rhizomes of, ginger, orris, and calamus.

Fruits of, bergamot, lemon, lime, and orange.

Seeds of, bitter almonds, anise (both kinds), fennel, and nutmeg.

Gums or *Oleo-resinous exudations* from, labdanum, myrrh, olibanum, Peru balsam, storax, and tolu.

Then again, different varieties of plants produce aromatic bodies of slightly dissimilar odour, as is shown by the numerous roses, such as the red rose, the white rose, and the Maréchal Niel, while yet again, the same plant, grown under different conditions and in different soil, will often yield an essential oil of entirely different bouquet, as is demonstrated by the lavender of Norfolk and of France or by the geranium of Vallauris and of Bourbon.

All these remarkable variations present a problem which has for many years been studied by numerous distinguished scientists, among whom may be mentioned Mer, Mesnard, Maquenne, Tschirch, Dr Eugene Charabot, and his co-workers, Messrs. Gatin, Hébert, and Laloue.

The theories advanced by some of these earlier workers concerning the formation of the essential oil in the plant is worthy of note.

Mer (1887) thought starches and cellulose were the starting-point in resin formation, preceded by that of essential oils.

Tschirch (1906), one of the greatest authorities on resins, agreed that the formation of the oil preceded that of the resins in the cell, but that they were produced from materials accumulated in the membrane of the cells bordering on the secreting canal.

Mesnard at an earlier date (1894), however, was of the opinion that essential oils were degradation products of chlorophyll. In the flower they are localised in the cells of the internal surface of the epidermis where by photosynthesis the chlorophyll is converted into essential oils, etc.

Maquenne more recently thought perseite and other polyhydric alcohols containing more than six-OH groups were the starting-point in the formation of aromatic terpenes.

All these workers considered the essential oils to be excretory products formed during the metabolism of substances which functioned in the life of the plant. They considered, further, that their property of odour had a distinct relationship to the functions of insects and of animals, but very little to that of life in the vegetable kingdom.

Charabot and Laloue, in the course of experiments conducted

over a number of years, were able to show that in many cases the essential oil did in fact originate in the chloroplast, and resulted from the assimilative work of the chlorophyll.

Concerning the formation of individual constituents of essential oils the following are the generally accepted views:

Alcohols formed first in the chloroplast.

Esters, by the action of acids on the alcohols in the chloroplast.

Hydrocarbons, by dehydration of alcohols in the chloroplast.

Terpene alcohols, by isomerization.

Acids, from the decomposition of proteins or from the oxidation of carbohydrates.

Aldehydes, from the rapid oxidation of alcohols principally in the inflorescence. Action hastened during fecundation and growth of fruit.

Ketones, probably in the same way as the aldehydes.

Phenols, either from the splitting up of proteins or of aromatic acids.

The problem of the evolution of these odoriferous constituents of the vegetable kingdom embraces the following points:

(*a*) The formation and circulation of the odoriferous constituents.
(*b*) Their evolution and the mechanism of that evolution.
(*c*) The creation of the perfumes themselves.
(*d*) Their physiological influence on the plant.

In a communication to the Academie d'Agriculture de France, Dr Charabot elaborated these points most clearly by considering the perfume first in the case of the whole plant and then in the case of the flower only.

He says: 'When the plant is examined the odoriferous materials only appear in the young organs and continue to form and accumulate with decreasing activity until blossom time; by diffusion they go from the leaf to the stalk and thence to the flower.

'During the process of fecundation a certain quantity of essential oil is consumed by the inflorescence, and as a practical consequence the gathering of perfume-producing plants should be made just before fertilisation is accomplished. Once this process is complete the fragrant principles redescend into the stem and diffuse into the other organs, that migration being stimulated by the drying of the inflorescences, which increases the osmotic pressure and partially precipitates the less soluble products.

'In considering the flower only, it is known that certain varieties (after collection) produce fragrant bodies when placed in such a condition that their vital functions may still be exercised, while in other cases the flower contains all its odoriferous principles in the free state, and it is impossible for it to produce new fragrant materials, even though it be still living.'

The conclusions arrived at by Dr Charabot after a study of the evolution of odoriferous compounds and of their mechanism are as follows: 'The esters, so frequently found in essential oils, are formed in a particularly active manner in the green part of the plant, by the action of acids on the alcohols. This phenomenon, characteristic of the chlorophyll region, is influenced by an agent, probably a diastase of reversible action, which functions as a dehydrating body. The influences capable of modifying the plant in order to adapt it to an intense chlorophyllic action at the same time aid the formation of esters, this being favourable to the mechanical elimination of water.

'Thus, the functions of chlorophyll tend to acquire a new significance; not only do they assure the fixation of carbonic acid gas by the vegetable tissues, not only do they, in assisting transpiration, effect the circulation of the liquids which bring and distribute the materials necessary to the mineral nutrition of the plant, but they also, during the assimilation of carbon, actively assist condensation, enabling the transformation of a simple chemical body into one of those innumerable complex substances, the study of which has puzzled the shrewdest chemist.

'When the alcohol is in the proper state to easily lose the elements of water, it gives birth to the esters and the corresponding hydrocarbon at the same time, or briefly put, the first transformation takes place in the chlorophyll region by way of dehydration.

'On the appearance of the flowers (those organs in which the fixation of oxygen by the tissues is particularly intense) it is possible that the alcohols and their esters are converted into other oxygenated products, the aldehydes or ketones, with at the same time the liberation of the energy necessary for fecundation.

'A large number of odoriferous materials, varying greatly in their functions and their chemical structure, may be produced by the splitting up of glucosides. When the generality of such a mechanism is admitted, it is easy to give an explanation of the observed facts relating to the formation of odoriferous materials and to their sudden appearance in this or that part of the vegetable organism. If the glucoside, which is formed in the green part of the plant, immediately encounters a medium whose conditions are favourable to its decomposition, the essential oil appears there at once and begins to circulate, to perform evolutions, and to play its part. In other cases, the glucoside will only meet the ferment capable of splitting it up in the flower. Only after having circulated in the plant and reached the flower, being modified more or less the whole way, will the glucoside be able to liberate the constituents of the essential oil. The flower only will then be odoriferous. The formation, in certain flowers, of new quantities of essential oil in proportion to the quantity of essence removed is explained by the phenomenon of chemical equilibrium, resulting from the reversible reaction:

$$\text{Glucoside-water} \rightleftharpoons \text{glucose-essential oil}$$

'The production of essence ceases when the state of equilibrium is attained. But, when the odoriferous material is removed in proportion to its formation, the reaction of division can go on until the whole of the glucoside is decomposed. From these conclusions it will be easy to understand their application in the extraction of flower perfumes, especially by enfleurage.'

With regard to the physiological influence of the aromatic materials, it was assumed formerly that they were of little use in the vegetable organism. It has been noticed by Dr Charabot and his co-workers, that, on the contrary, they can be employed by the plant, especially when the latter is sheltered from the light and does not assimilate the carbonic acid gas of the air with the same power. They participate in a normal manner in the work of fecundation and of the formation of the seeds, during which time they are partially consumed. Other hypotheses put forward to account for the part played by the essential oils in the life of the plant are as follows:

Ciamician and Ravenna think they may act as hormones, and thus act as excitants in the fecundation of the flower.

Pokorny considers them to be waste products because they are generally toxic both for lower organisms and for higher plants. Tschirch thinks they are waste products from which the resins are

THE PRODUCTION OF NATURAL PERFUMES 21

formed and are therefore incapable of circulation again in the plant.

Frisch takes the view that odour is more effective than colour for attracting pollen-laden insects, but that the essential oil has other functions is not disputed.

The suggestion that the volatile oil may be a protecting agent against plant parasites will not hold water because unfortunately odoriferous species are just as much subject to invasion as non-odorous plants.

Before leaving the subject of plant metabolism let us take a concrete example of the marvellous changes which occur in the composition of an essential oil—namely, that which is produced by the orange plant. If an essential oil is distilled from the inflorescences when they are in full flower and before fecundation has taken place, the product will have a comparatively high content of esters and other oxygenated bodies and be relatively low in terpenes.

If the orange flowers are fertilised and the fruit allowed to develop slightly, an essential oil is obtained on distillation that contains much less oxygenated constituents and a larger proportion of terpenes than the oil distilled from the fresh flowers.

Again, supposing the fruit is allowed to become fully grown and the essential oil expressed from the mature peel, it will be found that the oxygenated constituents have decreased to an almost negligible percentage and their place has been taken by terpenes.

Times of new crops

January.   Bois de rose, clove, linaloe, bergamot, lemon.
February.   Cassie, mimosa, bergamot, lemon.
March.   Violet, citronella, clove, Bourbon geranium.
April.   Jonquil, narcissus, hyacinth.
May.   Rose, orange blossom, rosemary, Bourbon vetivert, Algerian geranium, linaloe, thyme.
June.   Rose, carnation, petitgrain, cassie, bois de rose, vetivert, Bourbon geranium.
July.   Jasmin, French lavender, Algerian geranium, lime, rose, rosemary, thyme.
August.   Jasmin, tuberose, lavender (English and French), caraway, lemon-grass, English peppermint, Florentine orris.

September. Aniseed, jasmin, tuberose, Ceylon citronella, English lavender, palmarosa, English and American peppermint, spearmint, French geranium.

October. Jasmin, tuberose, caraway, spike lavender, Sicilian orange, palmarosa, French geranium, Florentine orris.

November. Cassie, Bourbon geranium, lemon, lime, American peppermint and spearmint, santal (Mysore auctions).

December. Cassie, bergamot, lemon, lemon-grass, palmarosa.

The separation of natural odoriferous materials

Production. The manufacturing processes employed may be conveniently classified as follows:

1. *Distillation* (geranium, lavender, neroli, rose, etc.).
2. *Expression* (citrus oils).
3. *Extraction* by means of (*a*) enfleurage, (*b*) maceration, (*c*) volatile solvents.

Distillation

This process in a crude form dates back to the times of the ancients, when fire was regarded as a supernatural element. Among the earliest types of apparatus were the *Cucurbita*, the *Alembic*, and the *Berchile*, from which has gradually been evolved the distilling apparatus of modern times. The process of evolution was slow up to the middle of the last century, but was hastened in later years, largely on account of the remarkable advances made in the perfume industry. During this period the improvement in the construction of steam and vacuum stills has not only increased and cheapened the yield of oil, but has materially enhanced the purity of the product. It does not, of course, follow that the employment of modern apparatus is universal, for even today, in more or less remote parts of the world, comparatively crude open fire stills are used, and their careless manipulation is frequently responsible for the oils of indifferent quality occasionally met with in commerce.

Volatile oils, as a general rule, are highly odorous, mobile liquids, and may be obtained from the plant by steam distillation without undergoing decomposition. They usually contain numerous individual bodies, differing in chemical constitution, and to one or more of these the characteristic odour of the oil is due. In many cases, these particular constituents have been definitely established, as, for example, the odour of almond oil is attributed principally to benzaldehyde, of bergamot oil to linalyl acetate, and of clove oil to eugenol. In several cases, however, where the oil is of more complex composition, the characteristic odour is believed to be due to the perfect blending of a number of the aromatic constituents. This is well illustrated in the case of otto of roses, where the esters of the alcohols geraniol and citronellol, together with the higher aliphatic aldehydes, although present in very small quantities, undoubtedly play an important part in the determination of the distinctive rose odour.

Different distillation processes are employed for separating the essence from the plant, and the choice of method depends upon the nature of the product and the yield that can be obtained. Several factors contribute towards the amount of essential oil that can be 'won' from the plant, and not least of these is the attention paid to its cultivation. In many cases, of course, the plant grows wild, and no effort has been made to study the particular kind of manure that could be used to materially increase the yield of oil. In a few instances, however, the importance of this part of the 'process' has been realised, and in the case of lavender proved by the experiments of Professor Zaccharewitz, of Avignon, who found that the yield of oil could be more than doubled by the use of an artificial manure consisting of sodium nitrate 1, potassium chloride 1, calcium superphosphate 3. Another important factor is the preparation of the plant for the still, and upon this will depend not only the yield of oil but also the rapidity with which the process may be completed. The state of the raw material must be such that the steam or water can completely permeate the mass and carry over with it into the condenser every particle of essential oil present. In many cases, such as flowers, leaves, and grasses, the raw material requires no special treatment, but sometimes the complete extraction of the volatile oil would be impossible without its previous preparation. In the case of hard, woody bodies, shaving or grinding may be necessary, in the case of some fresh roots or whole herbs they may be dried and then cut, while yet again in the case of seeds, fruits, and barks, crushing or disintegration may prove sufficient. In a few instances, notably berries, these preliminary preparations may be unnecessary because the epidermis is broken by internal pressure, induced by the high temperature of the steam. Raw materials are seldom placed in the still in a *fine* powder since they are apt to form an impenetrable mass when the steam is turned on. In consequence this escapes either round the sides or through cracks, and the major portion of the charge remains untreated.

In some parts it is the common practice to dry the raw material before distilling, but this may be due to the distance of the place of collection from the apparatus or to the congestion of the stills at the time; in the case of orris the drying of the roots is necessary for the development of their odour.

The processes of distillation may be divided broadly under the two following heads:

1. By boiling with water.
2. By means of steam ('dry or live').

The first method is the oldest and the easiest. It is still applied by the peasant who grows his own plants and is too far from the factory to convey them there. With careful application good results are obtained, but if a part of the raw material should not be covered with water and come in contact with the hot sides of the apparatus, destructive distillation takes place with the production of obnoxious bodies and a consequent impairment of the odour of the final product.

This last danger is overcome by the application of the second method, when the steam is produced in a boiler or other vessel away from the still. In this case the apparatus is fitted with a steam jacket, or a steam coil (for 'dry steam' distillation), and also with a 'live steam' inlet. Either system may then be used, and if necessary may be conducted under reduced pressure (*in vacuo*). The methods adopted for supporting the raw materials inside the still vary—in many cases a perforated sheet of metal or false bottom is used, while in others a basket of the same diameter as the still and fitting closely to its sides is suspended from the top and is easily removed.

The rate of distillation of the essential oil depends mainly on the condition of the raw material and the rate at which the volatile oil is liberated from it. This is influenced to some extent by its vapour pressure and the molecular weight of its constituents.

Condensation is effected by means of water-cooled coils or vertical tubes, the latter being very efficient and rapid for many products. If they are too fast, however, air cooling is resorted to as in the case of orris oil. To save the space and the cost of the enormous length of tube necessary for air cooling, worms can be inserted in warm water kept at any specific temperature by means of a thermostat and condensation efficiently effected.

Many ingenious forms of receivers are in use whereby the oil flows from one exit and the condensed water runs back into the still automatically. Florentine flasks are also used, both singly and in cascade. In the case of crude oils rectification is effected by redistillation with steam.

In a number of instances, notably in those of orange flower and rose, the oil, or some of its constituents are slightly soluble in the aqueous distillate, and even by subjecting the fragrant water to cohobation the complete separation of the volatile oil is difficult and frequently impossible. Such waters enter commerce as Aqua–Trip.

In some works these waters are specially prepared and are then known as 'weight-for-weight' products. For example, 100 lb of roses and a sufficient quantity of water are distilled until the yield is 100 lb of fragrant rose-water. Sometimes the distillate is separated into two or even three fractions; the first 50 lb constituting the quadruple superior, and the remainder the triple superior.

Expression
There are three main processes for the separation of so-called *Citrus* oils from the peel of the lemon, orange, bergamot, and lime. They may be described as:

(*a*) Sponge process.
(*b*) Ecuelle method.
(*c*) Machine process.

The first is applied to lemon and orange oils; the second was at one time employed for lemon oil but is now little used; the third on a large scale is applied mainly for the production of bergamot oil, but, as described below, machines have been devised to deal efficiently with lemon oil also. A small quantity of oil of limes is prepared by the sponge process, but as is well known the major portion of the oil of commerce is obtained as a by-product of distillation in the West Indies.

The sponge process
The oil cells of the rind of any of the above fruits are easily broken, as can be shown by turning a piece of lemon peel

backwards. This process on a large scale, therefore, does not offer any serious difficulty, nor does it require very heavy pressure for the extraction of the oil. It may be divided into three stages: (1) the preliminary preparation of the peel; (2) the expression of the oil; (3) the clarification of the oil. According to the manner in which the preliminary operation is carried out, so is the process named. When the fruit is cut across the shorter axis and the pulp removed by a spoon, it is known as the *Scorzetta*, and the sponges used are cup-shaped. When the rind is cut off in three strips and the pulp remains intact, it is known as the *Spugna*, and the sponges are flat or nearly so. This stage of the process is comparatively light work and is done by women. Before expression, the peel is either moistened with or steeped in water, which is supposed to facilitate the removal of the oil by making the cells more turgid. The drained peel is pressed by men who sit on low forms and allow the oil to collect in a shallow earthenware bowl in front of them, the sponges used for the purpose resting on sticks attached to the edges of the receiver. The oils from the different jars are mixed and allowed to stand until any juice has separated at the bottom. It is afterwards filtered and stored in coppers. Any residues that may contain oil are diluted with water and the oil recovered by distillation. Such products are always of poor quality and are therefore mixed with better oils. The hemispherical rinds from the *Scorzetta* process are pickled with salt and exported as *Salato*. The juice from the pulp is used in the production of *Citrates*.

The Ecuelle method

The Ecuelle method is practiced more in the north of Italy, and consists of rolling the fruits about in hollow vessels, the walls of which are covered with spikes. The oil cells are punctured, and the liquid flows to the bottom, being collected in a receptacle situated in the handle of the vessel. The product is then clarified as described above.

Machine processes

Machine processes have been applied in the manufacture of lemon and bergamot oils during recent years. One for the production of lemon oil consists of a mechanical adaptation of the sponge process, when the pressure is applied with a lever. Another hand

THE PRODUCTION OF NATURAL PERFUMES 27

machine consists of two channels between which the fruit is rolled; the skin is lacerated by means of spikes, and the mixture of oil and juice is collected and subsequently separated and clarified as above described. In another type of machine the fruits are freed from pulp and then the rind is placed between two rotative wire sieve-like plates. The oil is thus pressed out and collected. Imperfectly pressed peel is treated with sponges. Yet another process uses centrifugal separators. The whole fruit, without any previous preparation other than washing, is crushed in a mill between two pairs of rollers placed vertically, one below the other. In this process the essence-bearing cells of the peel are all broken and the essence thus liberated mingles with the juice simultaneously squeezed out from the pulp. The rollers are surrounded by a watertight casing, and the hopper of the mill is covered by a close-fitting, dome-shaped lid, the fruit being admitted by a shoot at one side, while from the middle of the cover a fine spray of water is directed downwards on to the first pair of rollers, and assists in washing the mixed juice and essence away from the residues. The mass is discharged from the lower part of the mill on to a wide, wooden grating, which retains the crushed fruit and allows the liquid, carrying in suspension a considerable amount of solid particles, to pass freely through it. On this grating the pulpy mass is allowed to drain for some time, and is then transferred to presses of the form usually employed in citrate works, where the remaining juice is squeezed out. The liquid passing through the grating runs down the sloping bottom of the tank and through a tube to the drainers, made of sheet aluminium pierced with tiny holes. In these strainers collect the pips and the coarser particles of skin or pulp that may be carried down by the liquid, which thus, partially clarified, passes into a tank from which it is led by a tube to the centrifugal machines. These are of the milk separator type, somewhat modified internally to permit of the easy passage of suspended solid particles to the wall of the drum, and to allow space for the accumulation there. The rate at which the liquid enters the drum is controlled by a stopcock in the tube, and when this is properly regulated, clear essence is continuously discharged from the upper or cream cock of the centrifuge, and collected in a suitable vessel, while juice freed from essence and containing but little matter in suspension is delivered from the lower cock and led by a pipe to an underground reservoir, whence it is pumped to the citrate-making plant as required. The centrifuges are employed in pairs, as after a certain period (one or two hours, depending on the state of the juice) the drum becomes overcharged with the deposit of finely divided solid matter, and must be dismounted and washed, the work being continued in the meanwhile by the other machine.

The machine process for the production of bergamot oil dates from the beginning of the eighteenth century. The somewhat crude apparatus was invented by Mr Auteri of Reggio, and consists of two circular-shaped discs, the lower one stationary and the upper one operated by geared wheels (either hand or power driven today). The discs are covered with either small knives or sharp points and the lower one is perforated to allow the oil to escape. The upper one is raised to insert the fruits and then lowered, the pressure being controlled by a lever. As the upper disc revolves the oil cells in the rind are lacerated and a bell rings to indicate when each batch is completely expressed. The oil escapes through the lower disc and is collected, being subsequently clarified and filtered. The fruits are afterwards cut up and the juice expressed for the production of citrates.

Extraction

Reference has already been made to the separation of volatile oils from plants (and their flowers) by means of distillation, and in several instances this process yields oils of exceptional purity and of very fine aroma. In a large number of cases, however, the application of this method does not yield products which are entirely satisfactory because many unstable aromatic substances are damaged or completely destroyed by the high temperature of steam, while in other cases the quantity of essential oil that could be obtained would be negligible. In view of these facts, other methods are used for the separation of the fragrant bodies from the flowers, and they are known as extractions by means of solvents. The materials used for this purpose are broadly classified as volatile and non-volatile, while the latter are again sub-divided according to the conditions of temperature appertaining during the process. These distinctions are clearly indicated as follows:

1. Extraction by means of non-volatile or fixed solvents such as animal fats or vegetable oils.

(*a*) At normal temperatures—Enfleurage.

THE PRODUCTION OF NATURAL PERFUMES 29

(b) With the application of heat—Maceration.

2. Extraction with volatile solvents such as petroleum ether, etc.

The choice of process depends upon several factors, the more important being:

(a) That certain varieties of flowers produce fragrant materials when placed in such a condition that their vital functions may still be exercised.

(b) That other varieties of flowers contain all their odoriferous principles in the free state, and are unable to produce *new* fragrant materials, even if still living.

Among the former class may be included jasmin and tuberose, while typical examples of the latter are rose and orange flower. The process which is best applied to the extraction of the odoriferous bodies from any particular flower has been determined by many years of experience in the south of France.

Enfleurage is applied principally to jasmin and tuberose and sometimes to orange blossom, jonquille, muguet, etc.

Maceration gives better results with cassie, rose, orange blossom, violet, etc.

Volatile Solvents are used for extracting reseda, rose, jasmin, jonquille, tuberose, violets, cassie, orange flowers, carnations, mimosa, heliotrope, oakmoss, stock, etc.

It will be noticed that certain flowers, such as rose and violet, may be extracted either by maceration or by volatile solvents, and a good product can be obtained by either process. On the other hand, certain flowers, notably jasmin, may be extracted by enfleurage or volatile solvents, but the product obtained by the former method is superior in odour rather than in yield. Niviere gave the following explanation of these differences:

'It is quite certain that the jasmin flower contains one or more glucosides. I had the opportunity of making a great number of experiments on the preliminary hydrolysis, and I always obtained a yield in essence greater than that obtained by a direct extraction without hydrolysis. However, if I compare the yields in essence by extraction with those by enfleurage, I notice that the differences are not very great, if only the extraction of *pure* essence is considered. During ordinary manufacture, 1000 kilos of jasmin flowers produce by enfleurage 5 to 6 kilos of pure concrete; extraction by means of petroleum ether only gives 1·300 to 1·400 kilos of absolute essence; the net yield accordingly seems to be four to five times greater by the enfleurage process. But it must not be forgotten that by this last method glycerides, rich in olein, are dissolved by the alcohol during extraction of the enfleurage pomade, and that in reality the jasmin concrete, from pomade, only contains 25 to 28 per cent of essence, the remainder consisting of fatty materials. What remains certain is that the two essences are different both in odour and in chemical composition, as demonstrated by Hesse. In the enfleurage product, the presence of indol is very noticeable, and is probably due to the splitting up of a glucoside under the influence of an enzyme. The fact of the vitality of the flower only intervenes in the direction of hydrolysis; the flower no longer produces essence, but splits up by hydrolysis the odourless or insoluble materials before they are absorbed by the fats. The fats dissolve the essence by contact. The industrial practice shows that the flowers must be in contact with the grease. All experiments to save labour, based on the non-contact of the flowers with the fat, in order to obtain a quicker 'defleurage', have failed. Concerning the use of these two essences in perfumery, it may be asserted that the enfleurage product is more tenacious than the absolute one. That peculiarity arises from the glycerides of the fatty acids, which act as fixatives, and furnish also their sweet odour of fat. The objections to the use of concretes is that they produce extracts which become rancid after some months, especially if the containers have been exposed to strong light.' This would appear to indicate that, if jasmin flowers were submitted to a preliminary hydrolysis by acid or enzyme before extraction with volatile solvents, the yield of oil would be increased.

Enfleurage

Enfleurage is the oldest perfumery process employed in the south of France. At one time it was used for extracting all blossoms, but modern methods have shown that better results can be obtained more economically with almost all flowers. The only important exceptions are jasmin and tuberose, and to these enfleurage possesses certain advantages dependent upon the fact that even after removal of the flowers from the stem, and while they are still in contact with the fat, the splitting up of glucosides continues with the production of essential oil which is absorbed and retained by the thin coating of grease. This fat yields the so-called

THE PRODUCTION OF NATURAL PERFUMES 31

'pomades for washings'. The enfleurage process necessitates the use of 'chassis', which are wooden frames, each supporting a glass plate. The cold or slightly warmed grease is uniformly distributed by means of brushes in a thin layer on both surfaces of the glass plate, a margin being left near the edges. The petals are then spread lightly on the fat, the absorption surface being increased by grooves made with a wooden spatula. Several chassis, thus prepared, are placed in tiers so that the petals are enclosed between two layers of grease, both the upper and lower absorbing the perfume as it is given off. Fresh flowers replace the exhausted ones, daily in the case of jasmin, and every two or three days in the case of tuberose. The chassis are turned over for each alternate application so that an even distribution of perfume results. The renewal of the flowers continues until the fat is fully charged, the resulting product, after removal, constituting the Pomade.

The fats most commonly used are lard or beef suet, or a mixture of the two in the proportion of two of the former to one of the other. They are specially purified and washed with alum solution, afterwards being preserved by digestion with benzoin or tolu (sometimes with the addition of orange blossoms).

The quality and cost of a flower pomade depend to a very large extent upon the care taken during their production and the cost of labour they have to support.

At one time, some Grasse manufacturers introduced a process whereby wire, silk, and yarn nets were placed upon the glass frame, between the layer of fat and jasmin flowers, as a perfume absorber. The results were not satisfactory.

This problem was solved by the design of a machine for mechanically removing the flowers. A frame, coated with jasmin flowers, is drawn into the machine, flower side downwards, and the flowers are blown off and fall to the floor. The frame still contains some remnants of flowers, which are too minute to be extracted so rapidly by the machine, regulated so as to 'snatch' the flowers as they pass by, without touching the grease coating in any way. The remnants of flowers are removed by means of a vacuum apparatus, the operation being exceedingly quick and the grease remaining untouched.

Occasionally solid fats are replaced with either olive oil or liquid paraffin, in which case cloths are saturated with the liquid and spread on chassis supporting wire frames, instead of glass plates.

When the oil is fully charged with perfume, the cloths are removed and submitted to hydraulic pressure. The perfumed oil so obtained is called Huile Française or Huile Antique. On extraction with one, two, or even three washings of alcohol, these products and the pomades yield the Extraits aux Fleurs, 1re, 2me, 3me, as the case may be. They are frequently sold as perfumes in this condition and without further treatment. The Concretes Soluble are obtained by removal of the alcohol from these *mixed* washings, the process being effected in a vacuum apparatus. In addition to the 'absolute' essence, they contain the glycerides of the fatty acids which are soluble in the alcoholic menstruum. The proportion of 'absolute' present varies, but as a general rule it is between 20 and 30 per cent of the total weight of 'concrete soluble'. These products are generally known as Enfleurage Absolutes.[1] The fat left behind after the various extractions is known as the Corps Epuisé and is used in the soap industry.

The flowers removed from the enfleurage greases are not entirely exhausted, and the remaining perfume content is extracted with petroleum ether by the usual methods. These products (of jasmin and tuberose) are known as absolute chassis. Owing to their slightly fatty smell they are much valued in the preparation of artificial flower oils since this characteristic facilitates the covering of the crude odour associated with most synthetics.

Maceration

Maceration consists in the extraction of the flowers by immersion in liquid fats or oils at a temperature of about 60° to 70°C. The greases or oils mentioned under enfleurage are used for this purpose, but paraffin appears to find less employment than the others on account of its lower absorption capacity. With the exception of jasmin and tuberose, all flowers, and particularly rose, are treated by this process. They are mixed with the hot greases in pans and the whole of the contents stirred. The cells containing the essential oil are ruptured by the heat and the aromatic constituent absorbed by the fat. When exhaustion is complete the contents of the pan are ejected on to a huge perforated screen and allowed to drain. The fat is collected and further quantities of flowers are mixed in with it, the process

---

[1] Compare also the monograph on Jasmin.

THE PRODUCTION OF NATURAL PERFUMES 33

being repeated until the extraction media is thoroughly saturated—the exact weight of flowers for the completion of this process having been arrived at by years of experience. The exhausted flowers left on the screen contain quantities of perfumed grease, and they are placed in linen bags and submitted to hydraulic pressure for its recovery.

Volatile solvents

This process was first experimented with by Robiquet in 1835. He extracted jonquille flowers with ether and obtained a concrete perfume of great delicacy. This method escaped the attention of chemists until 1856 when Millon extracted the perfume from numerous flowers by treatment with various solvents, such as chloroform, benzene, carbon disulphide, methyl and ethyl alcohols, etc. The yields were of good quality and the process was tried commercially, but owing to the loss of solvent it never paid and was discontinued. In 1879 Naudin patented a closed apparatus which eliminated these losses, and with the advent of petroleum light fractions the process offered commercial possibilities, but it was not until about 1890 that its real application on an industrial scale was attempted. Several solvents have been tried such as benzene, carbon disulphide, etc., but for various reasons they have been discarded, and the one in most general use at the present time is petroleum ether, sp. gr. 0·650 (15°C). This is first purified by treatment with sulphuric acid and alkali and subsequently rectified. Later benzole came into prominence because in the case of jasmin and rose a higher (and consequently cheaper) yield was obtained. The extractors resemble a battery of percolators, each having a capacity up to 100 gallons and fitted with several trays or perforated cylinders, in which the flowers are placed. The vessels are hermetically sealed, and connected the one with the other by means of tubes. At one end of the series they connect with the solvent tank and at the other with a vacuum still. The solvent runs through slowly and, when it reaches the vacuum still, is distilled off and returned to the solvent tank. The perfume remains behind in the retort. The solvent continues to pass through the apparatus until the flowers in each unit are exhausted in turn. They are then replaced without interfering with the continuity of the process. The product left behind in the retort is solid or nearly so. It contains the odoriferous materials together with the natural and insoluble plant waxes, and is known as Concrete (*parfum naturel solide*). Twenty-five grams of this extractive represents about one kilo of pomade No. 36. The yield per cent of flowers (weight) is as follows for the cases cited:

	per cent
Cassie	about 0·4
Jasmin	about 0·3
Rose	about 0·24
Violet	about 0·15
Orange blossom	about 0·282
Mignonette	about 0·14

In working the above process it is customary to treat the flowers as received from the fields. Messrs. Lautier Fils found, however, that by subjecting them to high pressure before treatment with volatile solvents, the yield of concrete is greatly increased.

The explanation given of this enhanced yield is, that the high pressure causes the cells to burst, thus allowing the glucoside and diastase to react more freely with the production of a greater amount of essence.

Another apparatus for the extraction of various odoriferous substances was patented by J. A. Hugues. In general structure it resembles a dredging machine. Two endless chains carry perforated boxes containing the flowers or substance to be extracted. One box at a time passes through the solvent, the whole being enclosed. When the solvent has become saturated, it is run off and separated in a vacuum still. The solvent remaining in the boxes is recovered by steam distillation.

Messrs. Lautier Fils patented the extraction method which necessitates the prior elimination of water from the flowers, etc., by the use of hygroscopic substances such as anhydrous sodium sulphate, etc.

In order to remove the plant waxes, the *concrete oils* are shaken with strong alcohol for twenty-four hours in machines called 'Batteuse', and the *insoluble* waxes removed by filtration. The alcoholic filtrate contains small quantities of *soluble* waxes, which are separated by freezing at about 20 degrees below zero. The Absolute flower oil[1] is obtained from this dewaxed alcoholic

---

[1] For the preparation of absolute chassis consult the monograph on Enfleurage in this chapter.

THE PRODUCTION OF NATURAL PERFUMES 35

solution, either by the removal of the solvent *in vacuo* or by the addition of salt, when the odoriferous essence separates on the surface and is collected. Standardised natural liquid absolutes are prepared from these essences by the addition of a neutral solvent and are completely soluble in alcohol. The yield of absolute flower oil (super-essence or specially purified) and of concrete by the volatile solvent process was recorded by Y. R. Naves, who obtained the following results, shown below, in the Laboratories of Messrs. Antoine Chiris.

In a number of cases the absolute flower oil contains varying quantities of colouring matter, which depends upon the volatile solvent used for extracting the flower. The presence of these

	Percentage Yield of	
	Concrete	Absolute from Concrete
Acacia	0·15–0·2	35–40
Broom	0·1–0·18	30–40
Carnation	0·2–0·25	9–12
Cassie—ancient	0·3–0·4	33
Cassie—Roman	0·35–0·5	33
Champaca	0·16–0·2	50
Gardenia	0·04–0·05	50
Hyacinth	0·9–1·05	60–70
Immortelle	0·6–0·95	45–50
Jasmin	0·15–0·2	10–14
Jonquille	0·28–0·33	45–53
Mimosa	0·3–0·55	40–55
Narcissus	0·25–0·45	27–32
Orange Blossom	0·2–0·4	36–55
Rose—Bulgarian	0·22–0·25	50–60
Rose—French	0·17–0·25	55–65
Tuberose	0·08–0·1	18–23
Violet—Parma	0·07–0·12	35–40
Violet—Victoria	0·08–0·18	35–40
Ylang-ylang	0·8–0·95	75–80

pigments is not always desirable when highly concentrated perfumes are being prepared, as they are liable to colour the handkerchief or dress to which they may be applied. In consequence, so-called Colourless Absolutes are in many instances manufactured, the separation of a portion of the pigment being effected either by steam distillation *in vacuo*, by co-distillation with ethylene glycol, or by exposure of the product to ultra-violet light rays. In the majority of cases, however, these products cannot be said to be completely colourless.

The following table shows the approximate weight of absolute flower oil that is required to produce one thousand of the type of perfume indicated:

	Strength of alcohol						
	90 per cent	80 per cent	70 per cent	60 per cent	50 per cent	40 per cent	30 per cent
Extract no. 72	25	—	—	—	—	—	—
Extract quadruple no. 36	22	20	18	12	—	—	—
Extract triple no. 24	15	13	11	9	7	—	—
Toilet waters	5	4.5	4	3	2	1.5	1

Note. These quantities are also applicable to *synthetics and terpeneless essential oils.*

Absolutes with synthetics

This process was evolved by E. Charabot. It permits the flower itself to complete a perfume mixture by keeping the latter in contact with the flower during the extraction of the natural perfume.

For instance, a lilac perfume is prepared as follows: the wax obtained by the extraction of jasmin flowers is added to a good quality lilac compound until the mixture has the consistency of enfleurage grease. The chassis are coated with this grease in the usual manner and then lilac flowers added, these being renewed daily in accordance with the standard methods of enfleurage. The flowers removed from the chassis are extracted and subsequently also the enfleurage fats. The two extracts are mixed and the absolute prepared by the usual methods.

Extraction with liquid CO_2

A German process has been patented in which it is claimed essential oils can be completely extracted without, in any way, impairing the odour of the natural product. The advantages are low temperature, quantitative yield, complete and spontaneous evaporation of the solvent from the extract.

Extraction with butane

The extraction of perfume from flowers by volatile solvents has been restricted, until recently, to the use of heptane (60/80 petrol ether) or pure benzene, with which are obtained the floral concretes.

The possibility of using homologues of heptane, although envisaged by Naudin in 1849, had never been seriously exploited.

The introduction of the use of butane opened a new era in the production of selected concretes from which may be obtained absolutes of very high quality called *Butaflors*.

This process was developed by Messrs. P. Robertet & Cie. of Grasse who succeeded after years of research in perfecting and patenting a process involving the extraction by butane of perfume from flowers, and in particular from certain delicate flowers such as Lilac, Muguet, Gardenia and Freesia, whose treatment by petrol ether had given unsatisfactory results.

The plant needed for butane treatment is much more complicated and costly than that used with solvents such as benzene and petrol ether, which are liquid at ordinary temperatures.

The employment of gas liquefied under pressure necessitates the use of leak-proof autoclaves capable of withstanding high internal pressures provided with control apparatus consisting of valves in special steel, storage cylinders and compressors, all of which are very costly.

In addition, the installation of an extraction unit requires a building made to precise standards whose electrical installations are made entirely of fire-proof material, together with the earthing leads indispensable for the removal of static electricity. Moreover, all the storage tanks are provided with *double-acting* safety valves.

The flowers are charged in extractors working under the pressure of 2 to 3 HPz and a countercurrent of liquid butane is allowed to circulate continuously until the flowers are exhausted.

The perfumed butane is collected in a special evaporator and leaves, on evaporation of the solvent at a temperature between $-20°C$ and $-10°C$, an odorous solid residue which is the concrete.

This concrete is treated in special pressure vessels. In a series of successive operations all the inodorous *waxes* are eliminated, and after evaporation of the solvent there remains an essence entirely soluble in alcohol, the *Butaflor*.

The average yield of concrete and of absolute based on five years of experience is as follows:

From 1000 kg. of treated flowers are obtained:

	Concrete	Absolute
Jasmin	2 kg 800	1 kg 400
Rose	1 kg 900	1 kg 150
Muguet	1 kg 100	0 kg 440
Tuberose	0 kg 700	0 kg 210
Lilac	0 kg 900	0 kg 180
Orange	1 kg 950	0 kg 980
Gardenia	1 kg 050	0 kg 280

The essential difference between ordinary absolutes and *Butaflors* is due to the fact that the perfume extraction takes place under pressure at a temperature below the temperature of the surroundings, and that the perfumed liquids are concentrated at temperatures below 10°C.

Conditions could not be more favourable for the retention of the full freshness of the flower perfume, and it may be said that *Butaflors* represent the nearest approach to an exact reproduction of the true flower perfume in all its delicacy.

Statistics

The average consumption of flowers at Grasse for the ten years 1948-57 and for the eleven years 1960-70 was as follows:

	48/57 tons	60/70 tons
Orange blossoms	700	422
Roses	500	318
Jasmin	700	364
Mimosa	100	197
Violet leaves	250	204
Narcissus	100	95
Genet	–	140

With the notable exception of Mimosa, which has always been relatively cheap, the decrease in the others is remarkable and reflects the increase in the cost of flower picking and delivery to the factories. The present prices for good quality absolutes has forced perfumers to cut their costs with synthetic flower oils in

THE PRODUCTION OF NATURAL PERFUMES

order to keep the sale prices of their products at an acceptable limit.

The relative importance of the foregoing processes in the treatment of flowers is approximately as follows:

Flowers	Distillation	Enfleurage and maceration	Volatile Solvents
Carnation	—	—	100
Clary sage	95	—	5
Hyacinth	—	—	100
Jasmin	—	10	90
Jonquille	—	10	90
Mimosa	—	—	100
Narcissus	—	—	100
Orange blossom	10	10	80
Rose	10	10	80
Tuberose	—	50	50
Violet leaves	—	20	80

Production seasons

Name	Principal districts of cultivation	When picked
Cassie	Cannes, Vallauris	September
Clary sage	Var, Vaucluse	June-July
Geranium	Grasse, Pegomas	October
Jasmin	Grasse, Pegomas, Mouans-Sartoux	July-October
Lavender	Alpes Maritime, Basses-Alpes, Vaucluse	July-September
Mimosa	Alpes Maritime, Var, Riviera	December-February
Orange blossom	Cannes, Le Cannet, Golfe Juan, Antibes	April-May
Peppermint	Pegomas, Villeneuve-Loubet	July-September
Roses	Grasse, La Colle, St. Paul, Vence, Le Bar	April-May
Rosemary	Gard, Herault	September
Sage	Grasse-Monteauroux	July-August
Spike	Riviera	August
Thyme	Riviera	April-June and October
Tuberose	Pegomas, Mouans-Sartoux	July-October

Yields per 100 kilos

Raw material	Grams
Almonds, bitter	220 to 240
Aniseed	1600 to 2000
Bergamot	660 to 700
Cassie	800 to 850
Cinnamon bark	450 to 1800
Clary sage	50 to 60
Cloves	16000 to 18000
Geranium	100 to 130
Laurel leaves	700 to 850
Lavender	800 to 1100
Mimosa	700 to 800
Mint (fresh)	700 to 720
Mint (dried)	2100 to 2800
Myrtle	250 to 300
Orange blossoms	700 to 1500
Orange peel	300 to 350
Origanum	500 to 760
Rose	50 to 80
Rosemary	500 to 550
Spike	500 to 1000
Thyme	80 to 120
Violet	3 to 4

The above figures must be taken as very approximate, since variations in flowers occur due to climate and soil.

CHAPTER THREE

The Purchase and Use of Flower Absolutes

When you buy flower absolutes, do you ask for an ordinary quotation?

If you do and want the best product, then you are making a mistake.

This statement on first sight may seem a little abnormal, especially so to keen buyers—and they exist in the perfumery trade. But in actual fact, as I shall endeavour to prove, this assertion will bear the closest inspection.

Rose and jasmin absolutes are the two flower products which sell big—the others, tuberose and orange blossom, have a smaller circle of users—while mimosa, cassie, etc., are only used by those perfumers who have experimented widely and found how to employ them very advantageously.

A flower absolute in the strict sense of the term means the whole of the perfume element in the flower (as far as it can be extracted) without natural waxes, pigments, etc. The volatile solvent process as is well known is universally adopted by the manufacturers of flower absolutes, and very probably over 90 per cent of them are produced by this method. The others (enfleurage absolutes) are (in a small percentage of cases of jasmin and tuberose) obtained by extracting the enfleurage pomade and in certain cases are to be preferred. The former, however, are generally asked for by buyers who know what they expect to get, so that for the purpose of quotations they are, more or less, obtaining a standard product prepared by a standard process.

It is necessary now to look a little farther back and examine the question of raw materials. It is customary for the flower growers to arrange the price of flowers before the crop is collected, and providing a reasonably good yield is anticipated after a normal season the different manufacturers pay the same price. If on the contrary atmospheric conditions have been such that a poor yield is anticipated, then it is often found that in spite of pre-arranged prices some speculator steps in and buys up as much of the crop as he can at a figure in excess of that agreed upon. Such instances come up now and again when there is naturally a slight variation in the cost of the finished absolute, but keen buyers observe these fluctuations which are recorded in the Grasse reports published in the Perfumery Press, and bear them in mind when making a purchase. When the crop is normal, however, no such fluctuations occur, but there is another aspect of the case which requires to be considered. Many of the larger firms grow a proportion of their more important raw materials. As progressive concerns they naturally study a question of soil, manure, situation, etc., with a view to obtaining blossoms rich in perfume and in consequence of high yield. It would, however, be absurd to assume that any one company has succeeded in consistently obtaining a better product and a higher yield than all other firms for every one of the flowers used by them in the production of absolutes. It can therefore be reasonably assumed that from the point of view of yield, all makers are using flowers which so nearly approximate that they may be considered to be of standard quality.

Having therefore agreed that the raw materials and process approximate so closely as to be identical, how is it possible to obtain any one flower absolute from two sources of supply—the one at sometimes double the price of the other? It is of course common knowledge that odourless or nearly odourless solvents, such as ethyl phthalate, benzyl benzoate, amyl benzoate, etc., are used as diluents, while some firms employ alcohol. Providing these dilutions are made clear to the purchaser no complaint can be made, but it is obviously wrong to send out dilutions described as absolutes. Manufacturers do not do this from choice but simply to meet competition—so called. However, today the more common method is to mix pure absolutes from various sources; for in the case of Jasmin these are available from Grasse, Algeria, Morocco, Egypt and Sicily, the cost declining in this order. Hence it is possible to buy a pure Jasmin Absolute containing, say, 10 per cent Grasse, 40 per cent Algerian and 50 per cent Sicilian.

When buying therefore ask your different agents to submit samples and prices for guaranteed pure absolutes. Insist on the fact that purity is essential and price of secondary consideration.

The difficulty next presents itself of which of the guaranteed pure absolutes to buy. The final test is of course that of odour. These delicate flower products are extremely unsatisfactory to judge in concentrated form, and a 1 per cent dilution in alcohol should therefore be made of each sample. On being left for a week and then a known number of drops placed on an odourless absorbent strip of paper a much better comparison can be made. Should there be any doubt between say two of the samples, it is more conclusive to make up specimens of the perfume for which the samples are intended and then make a comparison after maturing.

The amount of business transacted in artificial flower oils of good quality is considerable. The main difference between these and those of indifferent quality is that the former *may* contain flower absolutes in fairly large proportion and the latter either an insignificant quantity or often none at all. It may be assumed, therefore, that in buying artificial flower oils the price paid is generally according to quality and varies according to the percentage of natural perfume contained. In compounding these flower oils the aim of the chemist is always to approximate as closely as possible the characteristic flower odour without the natural perfume. Even though he may succeed in getting a fairly accurate reproduction of the odour with the skilful use of essential oils, synthetics, natural isolates, and in particular higher fatty aldehydes, there is always the smooth softness of the flower lacking. The pure flower absolute here supplies the missing link. It covers up those imperfections of odour as no other, at present known, substance. The more skilled the artist the less flower absolute he uses to obtain that soft finished otto.

There are several points in connection with the manufacture of artificial flower oils which must be borne in mind by the seeker after perfection. For instance, the percentage of flower absolute necessary is influenced by the ratio to one another of essential oil, terpeneless essential oil, natural isolate and synthetic used in imitating the flower odour note. It naturally follows that by using terpeneless oils the concentration and power of the resulting artificial oil is enhanced. Further, a terpeneless oil is generally preferable to a natural isolate, because by reason of the minute traces of known and also unidentified esters, ketones, aldehydes, etc., always present in the former, it has a softer odour note. Synthetics, on the other hand, are generally comparatively coarse in odour, but they can be softened appreciably by the addition of traces of clary sage oil or concrete. The odour notes of these two products differ—the former is more powerful and will go farther in consequence. The latter has a distinct resemblance to that of amber. All the higher fatty aldehydes from C_8 to C_{13} have an extensive use in finishing off artificial flower oils. The choice of the right member of the series is imperative, and great care is necessary so that an excess is not added. Too much fatty aldehyde will soon ruin any flower oil.

With the possible exception of jasmin and tuberose, the flower absolutes prepared by the volatile solvent process are used up to whatever percentage price will allow, and they yield excellent results. In the case of jasmin and tuberose, however, many chemists will have found distinct advantage with the use of enfleurage absolute. This is prepared by extracting jasmin pomade with petroleum ether. The absolute thus prepared has a more fatty odour than that prepared by the ordinary process, and in the production of the finished otto this shade of odour assists in covering the rough edges of the synthetics more effectively than the other. Having selected the best raw materials and blended them in the requisite proportions, the final softening of the artificial flower oil may be hastened by refluxing the whole at a warm temperature, using a water-bath as the source of heat.

In the production of alcoholic perfumes the soft odour note in all the best products is due to the perfect blending of the constituents and to a liberal use of flower absolutes. One of the reasons why well-known perfumes of English and French manufacture are so popular is that the makers use plenty of natural and comparatively little of synthetics; the latter are used to obtain the distinctive note and the former to enhance the bouquet.

In the preparation of face powders the perfume and colour are in nearly all cases the real selling features (especially the first sale). Flower absolutes used alone to perfume these products are an entire failure. Although the constituents of the powder are themselves odourless, they seem to alter the softness of the flower note and give it an odour which cannot be described better than of an "acid" nature. It is more than ever necessary in these products therefore to give the powder a warm soft fragrance as a basis on which to build the finished perfume. For this purpose the following substances are useful in the proportions indicated:

Heliotropin	18
Coumarin	5
Vanillin	2
Musk ambrette	5

The odour is then built up with essential oils and synthetics and the flower absolutes added to yield a soft elusive fragrance.

Many cosmetic creams placed on the market are perfumed with essential oils and synthetics only. When flower absolutes are used, great care must be exercised to ensure that the correct ones are chosen, otherwise discoloration of the cream results. The two which must be avoided are jasmin and orange blossom. These contain indole, and although the proportion of absolutes may be infinitesimal the cream nevertheless soon assumes a yellowish-grey appearance, and if exposed to sunlight this becomes reddish. When the jasmin odour is desired it is always safer to use rose absolute, amyl cinnamic aldehyde, and benzyl acetate, this combination having a similar bouquet in a compounded otto. In the case of orange blossom, it is advisable to employ rose absolute, linalol and terpeneless French petitgrain oil.

CHAPTER FOUR

Odour Classification and Fixation

Based on a paper read by the author in New York at the Annual Medal Award of the American Society of Cosmetic Chemists.

Perfumery has often been compared with music and with painting, in that creative work in all of them requires the artistic application of similar basic principles; if the fragrance, symphony or canvas is to receive general approbation, or even dire comment, from both critic and public.

There is a parallel in the original motif of each: for the musician builds up a series of notes to form his new theme; the painter discovers a beautiful scene and makes a rough sketch of his new subject; and the perfumer subtly blends certain tenacious aromatics to produce his new base. The elaboration of each idea demands a wide knowledge from the originator, a persistence of purpose and immense patience if the motif is to ultimately attain perfection.

The composer must understand both harmony and counterpoint, together with the special qualities of each instrument in the orchestra. He may use one or more of them to modulate his theme, and the others to impart a harmonious background; the whole producing an intricate tone poem in which he may vary his rhythm and key to retain and charm the ear of the listener.

The painter must be a master of colour blending, and especially so of the greys, if he is to not only effectively portray his subject, but also to impart atmosphere, lighting and perspective to his finished canvas.

The perfumer must have an intimate knowledge of perhaps a thousand aromatic substances, he must be familiar with the odour value of each, its source and the characters that determine its quality, as well as its floral type and blending range. Moreover, he must also select and use each one of them to skilfully compose, reinforce and shade his basic note until it becomes a symphony of fragrance. But, this is not all; for if his creation is to be employed as a perfume for cosmetics and soaps he must also have a profound knowledge of their composition and the effect each of their raw materials will have upon every constituent of his original fragrance, which involves cunning substitution and fresh blending to ultimately achieve the same finished note.

Of the three arts, it might be said that the appraisal of painting is the easiest, because the permanence of each canvas allows it to be more understandably appreciated and assessed by the eye, whereas the intangible nature of both music and perfumery defeats any concrete evaluation by either ear or nose.

Now, it is strange but true that the fundamentals of the three arts are not equally understood; for on both music and painting many detailed works have been published that can be studied by the potential composer or painter, and from which much guidance may be obtained in the course of their work. It is also true that research has provided a solid background to the work of the perfumer, in so far as the chemical and physical constants of each of his raw materials have been established, as well as the almost complete analysis of most of the essential oils, flower concretes and absolutes, all of which are available in many text books. Yet, despite this fund of information, it is unfortunate that none of it is of any real value in his creative work. I, myself, may have been guilty of offering, in my three works, suggestions and perhaps even novel ideas that the imaginative experimenter may employ and elaborate in the practice of his profession, but I freely admit that they are more useful to the novice than as a guide and friend to the expert.

When I wrote my first book on perfumery I was conscious of this defect, and although more information is available today on what I will call the 'Missing Link', and it is still being investigated by different chemists both here and in Europe, nothing of a practical nature had then been published about it. This link is a classification of odours and its application, and it was as long ago as 1926 that I began serious work upon it in an attempt to evolve a grouping of aromatic substances on a really sound basis. It took me no less than four years to complete, and you may well speculate upon the reasons why I did not publish the results in the next edition of my books. But at the time I felt it was too valuable for widespread diffusion, and so now it gives me pleasure to tell you about it here.

Before the introduction of synthetics the problem of classification was not quite such a difficult one, because the perfumer had at his command only the commoner essential oils and flower extracts. Chemistry has since evolved not only numerous shades of each odour type, but it has also created entirely new perfume bases. Moreover, many more essential oils are now available, and the whole gamut consists of perhaps a thousand raw materials against a former fifth of that number.

Rimmel in concluding his book of perfumes endeavoured to classify the then known substances by adopting a type for each class of odours and grouping with them other materials of similar fragrance. He was able to reduce his classification to eighteen distinct types which are given on this page.

Piesse in his art of perfumery took quite a different view of classification and compared odours with sounds. Scents, like

Classes	Types	Odours belonging to the same class
Almondy	Bitter almond	Laurels, peach kernels, mirbane
Amber	Ambergris	Oakmoss
Anise	Aniseed	Badiane, caraway, dill, fennel, coriander
Balsamic	Vanilla	Peru, tolu, benzoin, styrax, tonka
Camphoraceous	Camphor	Rosemary, patchouli
Caryophyllaceous	Clove	Carnation, clove pink
Citrine	Lemon	Bergamot, orange, cedrat, limes
Fruity	Pear	Apple, pineapple, quince
Jasmin	Jasmin	Lily of the valley
Lavender	Lavender	Spike, thyme, serpolet, marjoram
Minty	Peppermint	Spearmint, balm, rue, sage
Musky	Musk	Civet, ambrette seed, musk plant
Orange flower	Neroli	Acacia, syringa, orange leaves
Rosaceous	Rose	Geranium, sweetbrier, rhodium rosewood
Sandal	Sandalwood	Vetivert, cedarwood
Spicy	Cinnamon	Cassia, nutmeg, mace, pimento
Tuberose	Tuberose	Lily, jonquil, narcissus, hyacinth
Violet	Violet	Cassie, orris-root, mignonette

ODOUR CLASSIFICATION AND FIXATION

The gamut of odours as arranged by Piesse

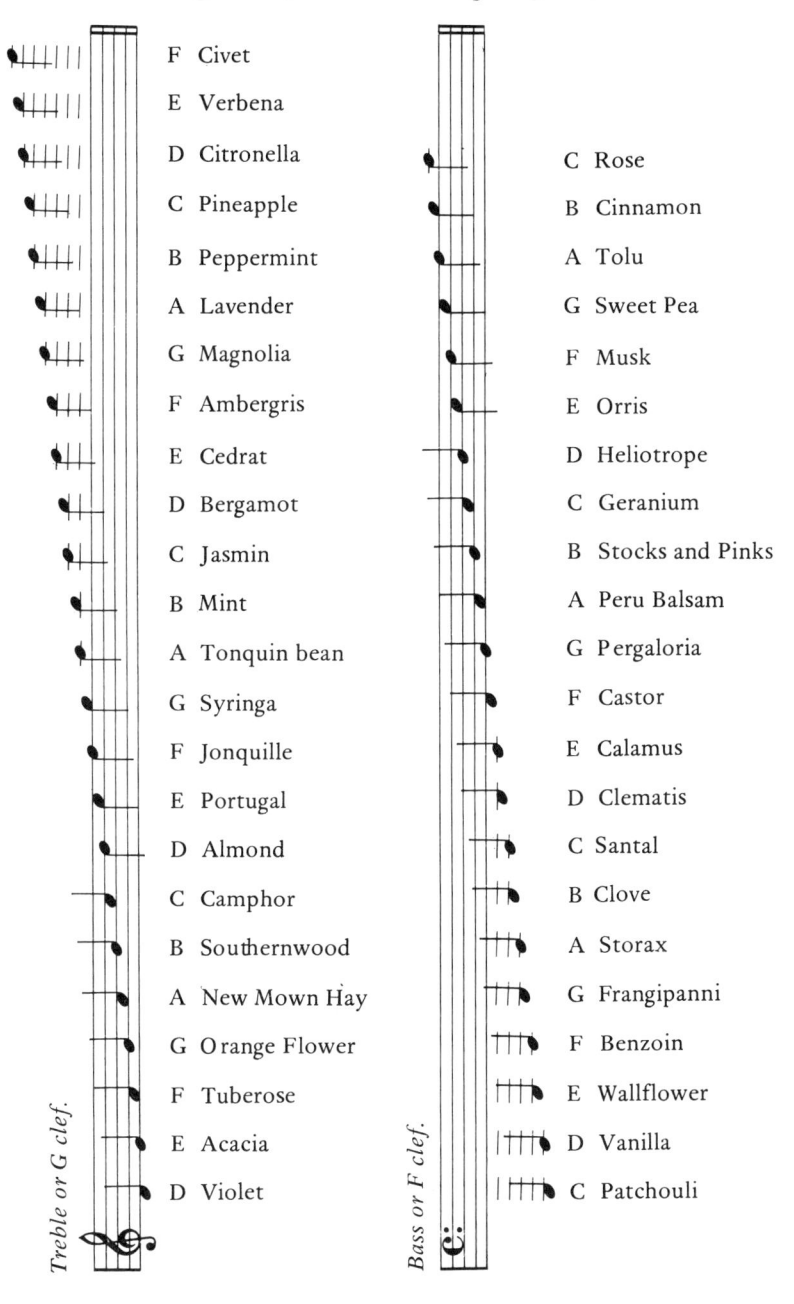

Treble or G clef:
- F Civet
- E Verbena
- D Citronella
- C Pineapple
- B Peppermint
- A Lavender
- G Magnolia
- F Ambergris
- E Cedrat
- D Bergamot
- C Jasmin
- B Mint
- A Tonquin bean
- G Syringa
- F Jonquille
- E Portugal
- D Almond
- C Camphor
- B Southernwood
- A New Mown Hay
- G Orange Flower
- F Tuberose
- E Acacia
- D Violet

Bass or F clef:
- C Rose
- B Cinnamon
- A Tolu
- G Sweet Pea
- F Musk
- E Orris
- D Heliotrope
- C Geranium
- B Stocks and Pinks
- A Peru Balsam
- G Pergaloria
- F Castor
- E Calamus
- D Clematis
- C Santal
- B Clove
- A Storax
- G Frangipanni
- F Benzoin
- E Wallflower
- D Vanilla
- C Patchouli

sounds, he says, appear to influence the olfactory nerves in certain definite degrees. There is, as it were, an octave of odours like an octave in music; certain odours coincide, like the keys of an instrument. Such as almond, heliotrope, vanilla, and clematis blend together, each producing different degrees of a nearly similar impression. Again there are citron, lemon, orange peel, and verbena forming a higher octave of smells, which blend in a similar manner. The analogy is completed by what are called semi-odours, such as rose and rose-geranium for the half-note; petitgrain, neroli, a black key followed by fleur d'orange. Then there are patchouli, sandalwood, and vetivert, and many others running into each other. In the gamut on page 49, Piesse endeavoured to place the name of the odour in its position corresponding to its effect on his olfactory sense. If a perfumer desires to make a bouquet from primitive odours, he must take such odours as chord together; the perfume will then be harmonious. In passing the eye down the gamut it will be seen what is a harmony and what is a discord of smells. As an artist would blend his colours, so must a perfumer blend his scents.

Crocker and Henderson[1] have made an analysis and classification of odours on a numerical basis. In their investigation they attempted to find the elements of sensation which make up all odours and came to the conclusion there are four kinds only. According to this point of view there are four types of olfactory nerves which are stimulated to differing degrees by the various chemical excitants which we call stinks, scents, or perfumes. These four apparently elementary odour sensations are:

1. *Fragrant* or sweet
2. *Acid* or sour
3. *Burnt* or empyreumatic
4. *Caprylic* or oenanthic.

These chemists arranged several hundred pure chemicals and some essential oils, according to the relative amounts of each odour component. To each component they assigned a figure or coefficient which noted the intensity of that component, as compared with a set of standards more or less arbitrarily assembled. They found it possible to easily note eight degrees and zero for each component.

[1] Amer. Perfum. (1927), 325.

The sensation 'fragrant' is prominent in the sweetness of flowers, in musk and violet ketones, and is noticeable in salicylates and cinnamates. It is practically absent in formic acid to which these chemists assigned the value 0. A moderate amount is noticeable in iso-propyl alcohol to which they assigned the value 3; decidedly sweet are citronellol and geraniol which they value at 6; and even more so iso-amyl and iso-butyl salicylates to which they gave the value 7: animal musk and some of the violet ketones are extremely sweet and were thus valued at 8.

The sensation 'acid' is that component of acetic and formic acids and sulphur dioxide which gives sharpness to the odour and which is popularly recognised as 'vinegary' or 'sour'. That it has remote connection with chemical acidity is indicated by its prominence in citral, camphor, acetone, and a number of other chemically neutral materials.

'Burnt' odour is well characterised and easily detected, for instance geraniol 2; citral 4; bornyl acetate 5; skatol 7.

'Caprylic' is a generally unpleasant sensation when strong, yet much missed when nearly absent. It is present in very small amount in good absolute alcohol, but in large amount in fusel oil.

Some of the standards adopted by Henderson and Crocker are as follows:

Fragrant

5222   Phenylethyl benzoate
6434   Diphenyl ether
7343   Safrole
8445   Benzyl acetate

Burnt

5414   Normal propyl alcohol
6322   Phenyl propyl alcohol
4376   Para-cresyl acetate
7584   Guaiacol

Acid

6123   Beta-naphthyl ethyl ether
5523   Iso-butyl phenylacetate
5636   Methyl phenylacetate
5726   Cineole

Caprylic

5221	Santalol
5322	Iso-amyl benzoate
2424	Toluene
2377	Anisol

The method which these chemists used in assigning odour numbers was to smell for but one component at a time. The following table gives the numerical limits of what can be easily recognised as the particular indicated type of odour:

Type	Fragrant	Acid	Burnt	Caprylic
Perfume	6-8	2-6	2-4	2-4
Lilac	7	4	3-4	3-4
Jasmin	6	3	3-4	3-4
Orange	6-8	5-7	1-3	2-4
Lemony	6-8	6-8	1-2	4-7
Spicy	3-8	4-7	any	2-6
Rose-like	5-7	4-6	2-3	3-5
Tea-like	4-6	0-4	1-3	1-5
Fruity	4-8	5-8	0-2	4-7
Mushroom-like	4-6	3-6	3	4-5
Musty	4-6	4-7	3-5	4-7
Sweet herb	4-8	4-7	any	5-8
Rank herb	1-3	4-7	1-5	5-8
Rum-like	4-6	5-8	0-2	5-8
Piney	4-6	5-7	any	3-6
Camphoric	4-6	5-7	4	3-6
Putrid	4-6	5-8	any	7-8
Greasy	3-5	6-7	5-7	5-7
Garlic and onion	3-5	6-8	3-5	7-8
Metallic	0-4	5-7	0-2	5-8
Tarry	any	any	7-8	6-8

If these three classifications were seriously considered by the modern perfumer, he would be bound to give the palm to Piesse; but I think it would be fair to say that he would find none of them of much use in the olfactory analysis of perfumes, or of their creation.

In more recent years further attention has been paid to this problem by different research workers, whose investigations have proceeded much on the same lines as my own of nearly thirty years ago; namely by grouping the synthetics in particular, together with some essential oils, according to their comparative volatility. But no attempt appears to have been made to give practical application to this valuable data, and later in this paper I shall attempt to elaborate my theories in such a way that they will offer some real assistance to the creative perfumer.

My first thoughts were directed to the molecular weights of the synthetics and natural isolates, but after tabulating them progressively I soon realised they could not reveal the secret of my quest. I next turned my attention to their boiling points at 760 mm, but even these, when tabulated, did not appear to solve my problem; whereas when taken at a pressure of 3 mm, which approximates more nearly the vapour pressure at room temperature, the answer seemed within my grasp.

However, the essential oils and flower extracts did not respond satisfactorily to this method, owing to their complex composition, and since they are the indispensable tools in the hands of every perfumery researcher I was compelled to adopt the only alternative, which consisted of the laborious estimation of their Duration of Evaporation by olfaction, at a laboratory temperature of 16°C. This was going to take much time and patience, but I felt it was the only method, which of course depended upon the reliability of my nose. Moreover, if I should succeed it might incidentally throw a brighter light on the dark problem of fixation. And here I must draw attention to a most important point; namely, the great improvement that has since taken place in the production, processing and purification of all aromatics, so that the results of my work may require adjustment and modification today. I hope that after the publication of this paper some interested perfumer will have the inclination and be prepared to give the time and patience that will be necessary in undertaking this work.

A prerequisite of this investigation was the careful selection of the finest and purest substances, both natural and synthetic, and this involved the repeated examination of all known aromatics from every available source. In view of the cleverness of sophistication, especially of the more expensive materials, this proved to be a long and trying piece of work. But I was fortunate enough to be able to obtain many of the natural products on the spot, in the various countries I visited, which thus guaranteed their purity.

Having selected and collected this vast array of priceless bottles, the next step was to decide on what should be the end point of the odour of the contents of each; for, as is well known, the characteristic note of some natural products may be fleeting while the residual smell lingers on. But since each aromatic substance is employed primarily for its typical odour note, I decided to check and re-check the point at which this distinguishing feature disappeared. Moreover, I had to place a time limit of those substances of longest duration, such as patchouli and oakmoss, and I gave them the figure or coefficient of 100.

My next step was the acquisition of a large stock of smelling strips, measuring 6 inches by $\frac{3}{8}$ inch, all made from the same stock of absorbent paper, and on each I wrote the name of the substance, the date and time of its application. It was important to have approximately equal quantities of the aromatics on the strips, but had I attempted to weigh exactly 100 mg of each, or calculated the differences due to their specific gravities, it would have added a further complication which I wished to avoid; and moreover, I felt it would not materially influence the time coefficient if I dipped quickly each strip to a depth of exactly one inch. Nevertheless, I made some preliminary tests to determine the differences, and while making no allowance for viscosity found the weights of the majority ranged between 80 and 100 mg, with an ultimate length of stain of about $2\frac{1}{2}$ in.

This method seemed reasonably satisfactory for the liquids and I adopted it, but I still had to find another method for dealing with the solids, which of course are invaluable and widely used. I therefore prepared 10 per cent solutions of them all in diethyl phthalate, and placed 1 cc of each on strips of paper measuring 6 in by 1 in, which gave me about the same quantity of each parent substance as the fluids.

As a guard against olfactic fatigue I limited the number of samples under examination to four a day, and began by smelling them every hour on the assumption that this figure or coefficient would be adequate. But when I found that most of them exhaled their characteristic odour for days on end I had to modify this view, and to those that evaporated in less than one day I gave the coefficient 1, and to the others 2 to 100. This work went on and on and soon seemed endless, because I had to make repeated tests when the results were doubtful. However, at the end of four years

I had finished, and when the substances were grouped under their respective coefficients, and the groups tabulated progressively from 1 to 100, I had before me a picture that was tantamount to the fractional distillation of any fragrance.

But I feel it is necessary to point out at this stage that the application of this key to the basis of perfumery will always depend upon the artistry of the worker in his laboratory, since without imagination it will never produce a masterpiece.

I am now going to give a selection from this classification of odours: the substances with coefficients from 1 to 14 I will call the Top Notes; those from 15 to 60 the Middle Notes; and the others from 61 to 100 the Basic Notes or Fixers, because they are fixatives in the usually accepted sense and aromatics which, when skilfully blended, yield the characteristic lasting note of any fragrance. However, I will return to the problem of Fixation later.

Top notes
1 Acetophenone, 2 Limes, 3 Coriander, 4 Lavender, 5 Terpinyl acetate, 6 Bergamot, 7 Geraniol, 8 Amyl salicylate, 9 Peppermint, 10 Linalyl acetate, 11 Sweet orange, 12 Methyl heptine carbonate, 13 Paracresyl phenylacetate, and 14 Lemongrass.

Middle notes
15 Heliotropin, 16 Eugenol, 17 Melissa, 18 Orris absolute, 19 Verbena, 20 Clary sage, 21 Anisic aldehyde, and the Ionones, 22 Orange flower water absolute, 23 Broom absolute, 24 Ylang, 31 Orange flower absolute, 34 Celery root, 43 Jasmin and rose absolutes, and 50 Neroly.

Basic notes
65 Cinnamic alcohol, 77 Methyl naphthyl ketone, 80 Hydroxy citronellal, 88 Ethyl methyl phenyl glycidate, 89 Cyclamen aldehyde, 90 Orris and opopanax resins, 91 Undecalactone, 94 Angelica root, and 100 Amyl cinnamic aldehyde, Coumarin, Decyl aldehyde, iso-eugenol, labdanum resin, Artificial musks, Oakmoss resin, Patchouli, Phenyl acetic aldehyde, Santal, Undecyl aldehyde, Vanillin, and Vetivert.

A question that arises from this classification is as to whether the constituents of one group may function temporarily in another if they are used in a dominating percentage in the finished compound. In this connection it is interesting to compare the vital constituents of lilac with those of hyacinth, because there is a similarity between them but with two substitutions and one notable omission. The following are common to both in this ratio: five times as much benzyl acetate in the hyacinth; six times as much phenyl ethyl alcohol in the lilac; the same percentages of heliotropin and of iso-eugenol in each; but the hyacinth requires three times as much cinnamic alcohol, and 100 times as much phenyl acetic aldehyde.

It will be observed that the almost equivalent changeover in the quantity of the two top note constituents is required by the difference in bouquet of the flowers, and is further modified of necessity by replacing the terpineol in the lilac with a similar quantity of the linalol in the hyacinth; the middle note of anisic aldehyde in the former is replaced by ionone in the latter; and while hydroxy citronellal is an essential constituent of the lilac, it can be omitted entirely from the hyacinth, which, however, is dominated by the vast increase in the powerful aldehyde.

It would seem, therefore, that longer lasting aromatics might be used in such quantity as to raise them temporarily into a higher grouping, as further instanced by the middle note of ylang in a lilac fantasy, where it might be used in such quantity as to dominate its fragrance; or the basic notes of such aliphatic aldehydes as C 10, 11, or 12 in an intense flowery bouquet, where they might be raised to middle, or even top, notes in accordance with the percentage employed. But in my view this cannot ultimately alter the original coefficients in this classification.

On the other hand, it seems certain that the relative quantities of aromatics in a complex mixture do in fact influence the duration of evaporation of the whole: first, depending upon the ratio of the combined basic substances to the middle and top note constituents; and second, to the changes that may take place on maturing, or even to interactions of an obscure nature between the various constituents. But it is not easy to prove the latter point, and an examination of the former may even lead to speculation. For, if we take a simple compound of two substances, such as bergamot and patchouli, or lavender and coumarin, and mix them in the proportions of 9 to 1, we shall find that although the odour of the major portion will be modified by that of the minor, the latter will last long after the former has evaporated. Furthermore, if we take an already blended perfume containing patchouli and coumarin, and substantially increase their percentages, we may find an overall increase in the lasting properties of the fragrance, but there would also be a shading of the note of the bouquet which might well spoil its lift and balance.

This brings me to the vexed question of fixation, to which I shall refer only briefly as it is incidental to the main theme. The problem of tenacity is as old as scents themselves, and although the ancient Egyptians and Grecians had few materials and no technical knowledge such as we possess today, yet they were able to make perfumes of lasting fragrance. This has been recently demonstrated by the discoveries in the tomb of Tutankhamen. The tenacity of a perfume was esteemed in those days, as will be clear on referring to the works of Theophrastus (370 B.C.). Concerning odours he says: 'Those perfumes whose scent is strongest get the best hold on the skin, head, and other parts of the body, and last the longest time: such are *megaleion*, Egyptian perfume, and sweet marjoram perfume. Those, on the other hand, which are weak and have not a powerful scent, since they are volatile and evaporate, also quickly come to an end: for instance, rose perfume and *kypros*. There are some, however, whose scent is even better on the second day, when any heavy quality that they possessed has evaporated. Some again are altogether more permanent, as spikenard and iris-perfume, and the stronger a perfume is the longer it lasts.' The same writer also refers to the properties of certain perfumes as follows: 'The lightest are rose-perfume and *kypros*, which seem to be the best suited to men, as also is lily perfume. The best for women are myrrh-oil, *megaleion*, the Egyptian, sweet marjoram, and spikenard: for these, owing to their strength and substantial character, do not easily evaporate and are not easily made to disperse, and *a lasting perfume is what women require.*' There are a few chapters in Pliny's works which treat of perfumes and unguents, but he does not appear to have had such a clear idea of fixation as Theophrastus although he lived some 400 years later and presumably in a more enlightened age. In that part of his work quoted in Chapter I, dealing with the Roman era, he speaks of resin and gum as fixators, and yet in Book XIII, Chapter 4, he says 'unguents lose their odour in an instant, and die away the very hour they are used. The very highest recommendation of them is, that when a female passes by, the odour which proceeds from her may possibly attract the attention of those even who till then are intent upon something else ... for the person who carries the perfume about with him is not the one, after all, who smells it.' I think we may safely assume that Theophrastus was the first to realise the implications of vapour pressure without knowing anything about it. Moreover, his womenfolk correctly assessed one of the most important qualities of a fragrance, which has not changed with the passing of the centuries.

In the first place, what is understood by fixation? Is it the retarding of the rate of evaporation of the more volatile substances in an already blended mixture of aromatics, by the addition of odourless, or nearly odourless substances of high boiling point; or is it the skilful blending of aromatics of low volatility, which not only largely contribute to the note of the perfume, but also inhibit the loss of the less tenacious middle and top note constituents? I think the tyro would accept the first premise, and the expert the second, mainly on the grounds that low vapour pressure solvents, such as diethyl phthalate and benzyl benzoate, have a distinct tendency to flatten the lift of a fragrance, and especially so if they are added in sufficient quantity to function as real fixatives.

You may, of course, regard this opinion as mere speculation, based upon long years of perhaps unconscious observation. So, with the sole object of proving, or disproving this belief, I made the following experiments. My first object was to discover which of the so-called 'fixatives' delayed evaporation for the longest period, and for this purpose I chose benzyl alcohol, b.p. 68°C at 3 mm; diethyl phthalate, b.p. 143°C at 3 mm; and benzyl benzoate, b.p. 147°C at 3 mm pressure: all of which are cheap and readily available. I would say in passing that I included benzyl alcohol because it is not an uncommon constituent of floral compounds. But these three by no means exhaust the list of usable substances, many of which I enumerated on page 86 in Volume II of the sixth edition of my works.

Nearly Odourless Synthetics with Boiling-points C

Amyl phthalate, 336°	Ethyl benzoate, 212°
Benzoic acid, 249°	Resorcinol diacetate, 278°
Benzyl benzoate, 323°	Tricresyl phosphate, 430°
Cyclohexanol oxalate, 220°	Triethylene glycol, 276°
Diethyl glycol, 245°	Triphenyl phosphate, 410°
Diethyl phthalate, 293°	

I selected two top note aromatics owing to their greater volatility: acetophenone, a straight synthetic, and bergamot, an essential oil of relatively complex composition; together with a first grade lilac compound containing the ingredients I shall shortly mention. To each I added 10 per cent of the 'fixatives', and then placed approximately 100 mg on smelling strips, making four in all of each aromatic.

The first smell clearly revealed the superiority of the pure substance, and by comparison the flattening effect of the additions; it was particularly noticeable in that containing the benzyl benzoate, which also imparted a rather metallic nuance. The strips were examined at regular intervals until the parent substance had evaporated, when it was clear that the odour strength of the other three were in direct ratio to the vapour pressures of the additions: thus, that containing benzyl benzoate was the strongest; diethyl phthalate less so; and benzyl alcohol the weakest. But the remarkable thing about these tests was that this rearrangement of strengths took place in one hour with the acetophenone; in six hours with the bergamot; and in fifty hours with the lilac.

Having therefore proved the superiority of the benzyl benzoate, I went a step further by adding it in the percentages of 1, 5, and 10 to anisic aldehyde and ylang ylang oil; aromatics of middle note according to the classification, the one of straight and the other of complex composition. In each example the pure substance was preferred on first smelling, followed by those of 1, 5, and 10 per cent additions. But since these two substances are inherently more lasting than acetophenone and bergamot, the weakening of their odours was less rapid; in fact, at the end of ten hours there was no change in the order of preference. It was only after fifty hours that a difference in the odour strength of the synthetic was noticeable, and then in the following order: 1 the 10 per cent addition, 2 the 5 per cent, 3 the pure substance and 4 the 1 per cent addition; whereas in the meantime the fragrance of the essential oil maintained its original order of preference.

Furthermore, I made parallel experiments with a Yardley perfumed cologne that contains many long-lasting ingredients. In order to decide whether the diluent of ethyl alcohol played a part, if any, in the duration of evaporation of the whole I made the same tests with the compound. The first smell confirmed the superiority of the 'unfixed' perfume and compound, and no change could be observed up to fifty hours, when the perfume containing 1 per cent of benzyl benzoate assumed first place, while those with 5 and 10 per cent additions still occupied third and fourth places respectively, but owing to the greater concentration of the compound the pure product retained first place after this time. Finally, I smelled the range of perfumes on the skin, when that containing no added fixative was predominantly fragrant.

It would appear, therefore, that these high boiling point odourless 'fixatives' exert a definite delaying action on the evaporation of the Top and Middle Note synthetics, but they do so at the expense of the freshness and lift of the pure substance. Moreover, they are not so active when added to essential oils by reason of their complex composition, since the low vapour pressure fractions probably act as natural fixatives to the more volatile constituents. In the same way the fragrance of a properly blended compound or perfume may be marred rather than enhanced by their use, and their employment should only be condoned as fixatives in basic compounds of a light, flowery or evanescent nature.

Hence, I think it is time we had a more precise conception of fixation as a whole. First, we should admit that it depends upon the choice and blending of the basic notes in any mixture of aromatics, and we might then with truth call this group of the classification fixatives if desired. Second, we should realise that no magical lasting qualities can be imparted by mere additions of odourless synthetics of high boiling point to already blended mixtures, and in consequence the perfumer will be bound to carefully consider their dual effect before adding them to his finished creation. For, if he is a master of his craft he will doubtless prefer to select his aromatics in accordance with their lasting properties, rather than rely upon dubious additions which might, in effect, impair his work of art.

To return to the main theme, how can this classification of odours be used to facilitate the work of the perfumer, either novice or expert? It may be true to say that the few are able to apply their genius exclusively to the creative side of the industry, but alas, the many are compelled to devote much of their time to the examination and duplication of other successful fragrances. This work can be exhausting and is often unsatisfactory, since to exactly copy a work of art is virtually impossible. Some of you will have experienced the disappointments that are inevitably associated with such problems; for after months of trial and error you may have succeeded in producing a colourable imitation of the model, only to find on offering it to your client that he expects to pay a quarter the price you ask for your laboriously produced compound.

This classification of odours can be of great assistance to this type of work; since it reveals to the eye, and suggests to the olfactory organ, all the possible constituents of the model in the order of their evaporation. This is rendered less difficult by placing a fresh quantity of the fragrance on paper at equal intervals and comparing the series of strips, when by difference the lost constituents will be immediately recognised by a keen and experienced sense of smell in accordance with the coefficients.

But it is in the creative side of perfumery that this classification can be so useful. The first step is to analyse the coefficients under the headings of the various flower types, and colognes, and to then tabulate them according to their respective groupings of Top, Middle and Basic notes. Each picture will then suggest the possible variations in shading as the new flowery creation develops, and trial and error by quantity will ultimately yield the required blend. I have abstracted the possible constituents for the creation of a lilac complex, from its simplest form in a compound to its more intricate pattern in a sophisticated fragrance, and have chosen this type because it is one of the commonest flowering agents in use. This particular classification is discussed later,[2] but in the meantime I shall quote from it the two examples I will discuss in detail.

In the creation of all flower compounds I give the same advice to my pupils, and will repeat it here and now; namely, that the key to success is *simplicity*, simplicity and again simplicity; or, in other words, the use of the absolute minimum of aromatics in the production of the finished blend. Bearing this in mind, an examination of the classification will reveal the vital constituents of a lilac as reducible to nine, as follows:

Top Note: Benzyl acetate 1, Terpineol 3, and Phenyl ethyl alcohol 4.
Middle Note: Heliotropin 15, and Anisic aldehyde 21.
Basic Note: Cinnamic alcohol 65, Hydroxy citronellal 80, Isoeugenol and Phenyl acetic aldehyde 100.

[2] See the monograph on Lilac.

The relative proportions of each are a matter of taste and I must leave each worker to compose his fragrance as he thinks fit, but if the perfume of pink lilac is to be duplicated with any accuracy, then attention should be paid to the predominating nuances of phenyl ethyl alcohol and hydroxy citronellal; their shading with heliotropin and cinnamic alcohol and crowning with terpineol, all of which are rounded off with relatively small percentages of the other essential ingredients, including benzyl acetate, which would be replaced by jasmin absolute in a compound of first quality.

This classification is perhaps even more useful in the creation of a flowery perfume, and in taking lilac again as the example, it will of course be based largely upon the compound already discussed. But, here *simplicity* gives place to *complexity*, as there are about eighty possible aromatics that may be used in building up a sophisticated fragrance of this type. Their selection offers endless permutations and combinations, and if ten perfumers worked on this theme every one of their creations would be different. Indeed, some might so modify the flower note that it would be unrecognisable as a lilac complex to all save the expert. I shall here discuss only a simple flowery version, which may be elaborated at will in accordance with the above ideas, as follows:

Top Notes: Bergamot 6 and Nerol 8.
Middle Notes: Ionone 21, Ylang 24, Jasmin and Rose absolutes 43, and Neroly 50.
Basic Notes: Cyclamen aldehyde 89, Santal, Musk ketone, Vanillin and Methyl nonyl acetic aldehyde 100; plus Lilac and Muguet compounds.

Finally, how can the mother chart be used in the creation of a new note? The most interesting, and even amusing procedure is to select any two substances of long evaporation, and to mix them in nine progressive proportions, as 1·9; 2·8; 3·7; 4·6; 5·5; 6·4; 7·3; 8·2; and 9·1. If the two chosen aromatics will blend then one of these mixtures will be a perfect balance, when neither can be detected by the nose as a separate entity, but when each contributes its odour quota to the new note. This simple procedure can be applied just as successfully to the flower compounds that are to be used to enhance the basic fragrance, and the note will be further reinforced and modified by suitable substances selected from the chart in accordance with their volatility, and in as cunning a manner as possible, which will of course depend upon the imagination of the perfumer.

Another procedure, in which the chart is only useful after the base has been created, is to employ the above arithmetical system to any two already compounded flower oils, such as violet and carnation. When their true blend has been found, the same arithmetical system is again used with this blend as one of the units, while the other should be a long lasting compound, such as an amber. This will yield a final basic blend of great and uncopyable complexity owing to the number of different constituents in the three compounds, and it can be further reinforced if desired in its middle and Top Notes by reference to the chart, and perhaps ultimately flowered with orange, rose and jasmin absolutes to yield a symphonic fragrance. In conclusion, the depth of such a creation merits the liberal use of the animal infusions of musk, ambergris and civet, which will impart life and diffusion to the whole.

Here, then, is an odour classification of practical value, which, in the hands of an experienced artist will not only reduce his time and labour by avoiding unnecessary experiments of a futile nature, but will also enable him to select with certainty the substances of proved value, knowing beforehand just what part they will play in his finished creation.

Duration of evaporation table

Top notes

1. Acetophenone
   Almonds
   Amyl acetate
   Benzaldehyde
   Benzyl acetate
   Ethyl acetate
   Ethyl acetoacetate
   Iso-butyl acetate
   Methyl benzoate
   Niaouli

2. Benzyl formate
   Bois de rose
   Ethyl benzoate
   Limes distilled
   Linalol
   Mandarin
   Methyl salicylate
   Octyl acetate
   Phenylethyl acetate
   Phenylethyl formate
   Phenylethyl propionate
   Phenylethyl salicylate

3. Benzyl cinnamate
   Coriander
   Para cresyl methyl ether
   Para cresyl acetate
   Para cresyl iso-butyrate
   Cuminic aldehyde
   Cyclohexanyl butyrate
   Decyl formate
   Dimethyl benzyl carbinol
   Dimethyl benzyl acetate
   Ethyl decine carbonate
   Ethyl salicylate

Top notes—cont.

Methyl acetophenone
Musk, 3 per cent
Myrrh oil
Octyl iso-butyrate
Pennyroyal
Petitgrain para
Sassafras
Spearmint
Terpineol

4. Dimethyl octanol
   Cummin
   Citronellol
   Eucalyptus
   Geranyl benzoate
   Lavender
   Methyl butyrate
   Myrtle
   Nonyl aldehyde
   Phenylethyl alcohol
   Sage
   Methylmethyl salicylate

5. Dimethyl acetophenone
   Ethyl phenylacetate
   Neroli, Italian
   Nonyl acetate
   Terpinyl acetate
   Para tolyl aldehyde

6. Bay
   Bergamot
   Caraway
   Cedrat
   Citronellyl formate
   Copaiba oil
   Iso-butyl phenylacetate
   Linalyl benzoate
   Phenylethyl benzoate
   Phenylmethyl carbinyl acetate
   Grape fruit

7. Amyl propionate
   Aniseed
   Benzyl iso-butyrate
   Benzyl propionate

Ethyl heptoate
Geraniol, Java
Ginger
Methyl octine carbonate
Nonyl alcohol
Pansy
Peppermint, Japanese
Rue
Tansy
Thyme white
Violet absolute, 10 per cent
Geranyl iso-butyrate
Decyl acetate

8. Amyl salicylate
   Benzyl salicylate
   Cedarwood
   Citronella, Ceylon
   Citronellyl acetate
   Ethyl anthranilate
   Geraniol ex palmarosa
   Iso-butyl benzoate
   Iso-butyl salicylate
   Lemon
   Linalyl propionate
   Nerol
   Rhodinol
   Rose otto, French
   Wormseed
   Phenylethyl valerianate

9. Dimethyl nonenol
   Geranyl butyrate
   Iva
   Laurel leaf
   Methyl anisate
   Neryl acetate
   Peppermint, American
   Spike lavender
   Tagette
   Thyme red

10. Absinthe
    Camomile
    Diphenyl methane
    Diphenyl oxide

ODOUR CLASSIFICATION AND FIXATION

Top notes—cont.

Ethyl anisate
Lavandin
Linalyl acetate
Methyl phenylacetaldehyde
Orange bitter
Phenylpropyl propionate
Methyl eugenol

11. Amyl butyrate
    Carrot seed
    Cubebs
    Decyl alcohol
    Galbanum oil
    Hyacinthe absolute
    Immortelle absolute incolore
    Kuromoji
    Linalyl cinnamate
    Linalyl formate
    Lovage
    Matico
    Narcissus absolute
    Nutmeg
    Octyl alcohol
    Opoponax oil
    Orange sweet

12. Methyl cinnamate
    Methyl heptine carbonate
    Petitgrain, French
    Phenylethyl cinnamate
    Terpinyl propionate
    Amyl benzyl ether

13. Paracresyl phenylacetate
    Elemi oil
    Orris concrete
    Phenylpropyl alcohol

14. Basilic
    Cananga
    Fennel
    Lemongrass
    Mastic oil
    Methyl ionone
    Mimosa absolute
    Palmarosa
    Phenylpropyl acetate
    Phenylpropyl iso-butyrate
    Reseda absolute

Middle notes

15. Acet anisol
    Cinnamyl acetate
    Cinnamyl formate
    Citronella, Java
    Dill
    Guaiac wood
    Heliotropin
    Skatole
    Styrax oil
    Rose otto, Bulgarian

16. Amyl anisate
    Eugenol
    Phenylpropyl phenylacetate
    Serpolet

17. Melissa
    Tetrahydro geraniol

18. Calamus
    Marjoram
    Orris absolute
    Phenoxyethyl iso-butyrate
    Styrolyl valerianate
    Violet leaves absolute

19. Phenylethyl iso-butyrate
    Verbena genuine

20. Clary sage

21. Amyl Benzoate
    Angelica seed
    Anisaldehyde
    Arnica root
    Elemi resin
    Indole
    Ionone, alpha and beta

Middle notes—cont.

      Methyl anthranilate
      Myrrh resin
      Rosemary, French
      Undecylic acetate

22. Benzyl iso-eugenol
      Cinnamon leaf
      Cinnamyl propionate
      Cloves
      Geranyl formate
      Linalyl anthranilate
      Orange flower water absolute
      Phenyl cresyl oxide

23. Broom absolute
      Methoxy acetophenone
      Parsley

24. Anisyl acetate
      Auracaria oil
      Benzylidene acetone
      Cinnamon bark
      Ethyl cinnamate
      Ethyl furfurhydracrylate
      Geranium, African, French and Spanish
      Geranyl acetate
      Jonquille absolute
      Ylang ylang, Manilla

25. Methyl iso-eugenol

26. Eucalyptus citriodora
      Iso-butyl methyl anthranilate
      Methyl phenylacetate

27. Citronellyl benzoate
      Dimethyl hydroquinone

28. Cinnamyl butyrate

29. Cascarilla
      Geranium, Bourbon

30. Ambrette seed oil
      Cardamon

      Gingergrass
      Limette

31. Orange flower absolute

32. Ethyl laurinate
      P. Methyl hydro cinnamic aldehyde

33. Zdravets

34. Celery root

35. Dimethyl anthranilate

38. Hops

40. Hyssop
      Rhodinyl propionate
      Ylang ylang, Bourbon

41. Mace

42. Acetyl iso-eugenol
      Amyl cinnamate
      Cinnamylidene methyl carbinol
      Gerindol

43. Eugenyl formate
      Iso-butyl cinnamate
      Jasmin absolute
      Rose absolute
      Tuberose absolute

45. Bornyl acetate
      Cassia

47. Anisic alcohol

50. Laurinic alcohol
      Laurinic aldehyde
      Phenylpropyl aldehyde
      Undecylic alcohol
      Neroli bigarade

Middle notes—cont.

54. Cedryl acetate

55. Nerolidol

60. Benzyl phenylacetate
    Citral
    Rhodinyl formate

62. Amyl phenylacetate

65. Cinnamic alcohol nat.

70. Linalyl salicylate
    Jasmin decolore

Bases

73. Cassie absolute, Farnesiana

77. Methyl naphthyl ketone

79. Civette absolute

80. Hydroxy citronellal

85. Phenylacetaldehyde dimethyl acetal

87. Octyl aldehyde

88. Ethyl methyl phenyl glycidate

89. Cyclamen aldehyde

90. Galbanum resin
    Opoponax resin
    Orris oleo resin
    Rhodinyl acetate
    Santal W.A.
    Tarragon
    Jasmin chassis incolore

91. Phenylethyl phenylacetate
    Undecalactone

94. Angelica root
    Birch bud

96. Arnica flowers

100. Acet eugenol
    Ambergris extract, 3 per cent
    Amyl cinnamic aldehyde
    Amyloxy iso-eugenol

Benzoin
Benzophenone
Birch tar
Castoreum absolute
Cinnamic alcohol synthetic
Costus
Coumarin
Cypress
Decyl aldehyde
Ethyl vanillin
Gamma nonyl lactone
Guaiyl esters
Immortelle absolute
Iso-eugenol
Iso-eugenol phenylacetate
Labdanum
Linalyl phenylacetate
Methyl nonyl acetaldehyde
Musks artificial
Oakmoss
Olibanum oil and resin
Patchouli
Pepper
Peru balsam
Phenylacetic acid
Phenylacetic aldehyde
Pimento
Rhodinyl phenylacetate
Santalwood E.1.
Storax resin
Santalyl phenylacetate
Tolu balsam
Tonka resinoid
Trichlor phenyl methyl carbinyl acetate
Undecylic aldehyde
Vanillin
Vetivert

CHAPTER FIVE

Monographs on Flower Perfumes

Acacia

Acacia is the name of an extensive genus of trees and shrubs of the N.O. Leguminosae, varying in habit from heathlike shrubs to lofty trees, and widely spread throughout the tropical and sub-tropical regions of both hemispheres. The inflorescences take the form of compact globose heads or spikes of various colours, generally white, pink, or yellow, the latter being the predominant colour in the Australian species.

Varieties

In India the genus is represented by about eighteen species of trees of various sizes, distributed throughout the country, some attaining a height of 100 feet, especially in the forests of Pegu and Prome. In Western Asia and Africa the genus is represented by gum-yielding species, such as *A. Arabica* and *A. Senegal*, which are small sized, thorny trees of forbidding aspect and frequently occupying large tracts of desert country. Other species of acacia are also common in the West Indies and tropical America, where they are valued for their timber. In Australia this genus is profuse, and as many as 300 different species are recorded, several of which are of great commercial value as the bark is used for tanning. Among the more important of these are *Acacia decurrens*, known as the black wattle in Victoria and Tasmania; *A. dealbata*,[1] the

---

[1] See also Mimosa.

silver wattle; and *A. pycnantha*, the broad-leaf wattle. In southern Europe and western Syria the genus is represented by *A. Julibrissin* and *A. Farnesiana*.² In this country, the south of France and northern America, the trees generally but erroneously referred to as acacia are *Robinia pseudacacia*, having aromatic white flowers, which appear during May and June and impart a pleasant odour to the avenues and gardens they adorn. Originally a North American species, the tree was introduced into Britain some 250 years ago and is today much admired. It attains a height of 40 or 50 feet and averages 2½ feet in diameter.

Odour

Several acacias possess sweet-scented flowers, and of these (with the exception of *A. Farnesiana* and *A. dealbata*, which will be dealt with in separate monographs) the more important are *A. biflora* and *A. hastulata*. The odour of the former recalls the coconut, while that of the latter resembles hawthorn, but as far as is known neither have been turned to practical account for the extraction of their perfume in Europe. In Australia, however, there is a perfume known as 'wattle blossom', obtained from the flowers, collected after sundown. It is prepared by macerating them in olive oil, which when saturated is extracted with strong alcohol. The odour of the flowers of *Robinia* approximates more nearly to that of the *A. hastulata* and will therefore be taken as the standard flower for its synthetic prototype.

Chemistry

The flowers of *Robinia pseudacacia* have been subjected to an examination by F. Elze, who extracted the blossoms with a readily volatile solvent and obtained a very dark-coloured oil with a peculiar basic odour which, when diluted, reproduced the natural flower fragrance. This oil contained 9 per cent of ester calculated as methyl anthranilate. In alcoholic solution it gave a clearly perceptible blue fluorescence, and on dilution with ethereal sulphuric acid yielded this substance. The following further constituents were also identified: indole, heliotropin, benzyl alcohol, linalol, and α-terpineol. In addition, aldehydes and ketones with a decided odour of peach, and probably also nerol are present.

² See also Cassie.

Odour classification

Top notes	Middle notes	Basic notes
1. Benzyl acetate	15. Acet anisol	65. Cinnamic alcohol
2. Linalol	Heliotropin	77. Methyl Naphthyl ketone
Rosewood	21. Anisic aldehyde	
3. Methyl acetophenone	Ionone alpha	79. Civet absolute
Terpineol	Methyl anthranilate	80. Hydroxy citronellal
4. Citronellol	22. Clove	88. Ethyl methyl phenyl glycidate
Nonyl aldehyde	24. Ylang	
Phenyl ethyl alcohol	31. Orange flower absolute	91. Undecalactone
6. Bergamot		100. Benzoin resin
Iso-butyl phenylacetate	43. Jasmin absolute	Coumarin
	Rose absolute	Iso-eugenol
8. Iso-butyl benzoate	50. Neroly	Musk ketone
Rhodinol		Peru balsam
12. Methyl cinnamate		Phenyl acetic acid
Petitgrain, French		Phenyl acetic aldehyde
		Santal
		Tolu balsam
		Vanillin
		Vetivert

Compounding Notes

Acacia perfumes, as distinct from those of cassie and mimosa, are characterised by an intense flowery fragrance that is reminiscent of a blend of hawthorn with orange blossom. A rich bouquet may be obtained by combining these already compounded oils, or if a basic note is required on which to build the perfume it may be secured by mixing anisic aldehyde with methyl anthranilate in the ratio of 4 to 6. But since the latter is rather harsh a part of it may be replaced with advantage by methyl naphthyl ketone, in which case the note may be rounded off by the following vital constituents:

    Top notes. Benzyl acetate, Linalol, Terpineol, Bergamot, Methyl cinnamate and French petitgrain.
    Middle notes. Clove, Ylang, Jasmin, Rose, and Neroly.
    Basic notes. Musk ketone, Santal, and Vetivert.

    Two formulae are appended: the first is of basic construction and the second more complex.

Acacia, no. 1001

10	Benzyl acetate
20	Linalol
30	Terpineol
20	Bergamot
20	Methyl cinnamate
50	Petitgrain, French
280	Anisic aldehyde
350	Methyl anthranilate
20	Clove
30	Ylang
10	Jasmin absolute
20	Rose absolute
10	Neroly
70	Methyl naphthyl ketone
30	Musk ketone
20	Santal
10	Vetivert
1000	

Acacia, no. 1002

50	Bergamot
20	Methyl cinnamate
280	May blossom compound
40	Ylang
420	Orange blossom compound
30	Jasmin compound
70	Rose compound
10	Iso-eugenol
30	Musk ketone
50	Santal
1000	

Handkerchief perfumes can be made from either of the foregoing as follows:

no. 1003

100	Acacia, no. 1002
30	Musk extract, 3 per cent
870	Alcohol
1000	

no. 1004

75	Acacia, no. 1001	
20	Civet extract, 3 per cent	
5	Jasmin absolute	
10	Rose absolute	
10	Benzoin resin	
880	Alcohol	
1000		

Carnation

History

The Greek philosopher Theophrastus, in his *Enquiry into Plants*, VI, 6, 2, states that the gillyflower (? stock) is sweet-scented, but that the carnation and wallflower are *scentless*, from which it is evident that this flower was known in the fourth century B.C. There appears to be no clear record of the introduction of the carnation into Britain, some writers stating that it came from Germany, and others that it was imported from Italy and the shores of the Mediterranean. There is no doubt, however, that the spicy fragrance of the flower has been appreciated for centuries throughout Europe, and was very much favoured in the time of Queen Elizabeth. In later years the poet William Shenstone wrote of it as follows:

> Let your admired carnation own,
> Not all for needful use alone;
> There while the seeds of future blossoms dwell,
> 'Tis coloured for the sight, perfumed to please the smell.

The clove pink, *Dianthus caryophyllus*, of which the carnation is a variety, is a grass-leaved herbaceous plant of the N.O. Caryophyllaceae. The origin of the name 'clove' is worthy of mention. It is derived from the French word *clou*, English 'clout', a nail, from the imaginary resemblance of the clove flower to the head of a nail. Tournefort, a French botanist who died in 1708, is supposed to have given it the specific name *Caryophyllus* on account of its similarity to some of the short-leaved species of the genus *Carex* and its allies. The word *Caryophyllus* is also applied to the molucca clove, although there is no likeness between the two; nevertheless Tournefort's name of *Caryophyllus aromaticus* was adopted by early botanists for this well-known spice. This consists of the dried, unexpanded flower-buds of a tree of the N.O. Myrtaceae, which is now known as *Eugenia caryophyllata*.

Varieties

There are over fifty species of *Dianthus*, with numerous varieties, believed by some horticulturists to exceed 2000, and including carnations, pinks, picotees, and sweet-williams. Some of the well-defined species are:

D. Chinensis—beautiful but inodorous.
D. Barbatus—the sweet-william.
D. Hortensis—the garden pink.
D. Plumarius—the pheasant's eye.
D. Deltoides and D. Caesius—both commonly occur wild.

Carnations are grouped by florists, according to markings of the flower, as follows:

Bizarres—spotted or striped with several shades (usually three).
Fancies—with markings on coloured grounds.
Flakes—of two colours, striped longitudinally.
Picotees—with tinted petal edges.
Selfs—of one colour.

On referring to the catalogues of any of the well-known horticulturists there will be found hundreds of these varieties of carnations, and, like roses, they are known by all sorts of fancy names. From the point of view of odour, however, white carnations are generally to be preferred to red ones. Along the French and Italian Rivieras very large tracts of land are devoted to the cultivation of these exquisite flowers. Visitors will have noticed them in particular near Nice and Antibes and also near Ventimiglia, Bordighera, and San Remo. The blossoms begin to appear as early as September and continue until July. The major portion of them are sold as cut flowers and sent to Paris and London. As an indication of the importance of this business, it is interesting to note that a substantial number of carnations are sold annually by French growers and a lesser amount by the Italians.

Odour

The carnation has developed its rich, spicy odour with cultivation, although it is a peculiar fact that horticulture is responsible for many beautiful forms which are almost devoid of perfume. In the wild state it seldom possesses either of these qualities, and is occasionally found growing on dry soil.

Natural perfume

Although, as stated above, large quantities of carnations are grown in the south of France, by far the greater proportion are sold for decorative purposes. In certain parts of the Var near Grasse the flowers are grown especially for perfumery purposes. Those of importance are white, pink, pink and red, and yellow and red. The harvest takes place in June, and the blossoms are picked after exposure to about three hours of brilliant sunshine. The perfume is then at its maximum fragrance. It is extracted nowadays almost exclusively by means of volatile solvents. About 500 kilos of flowers yield 1 kilo of concrete. This has a rather high wax content and the yield of absolute is in the region of 10 per cent only. This has a waxy odour of heavy carnation type and is eminently suited for sophisticated perfumes. In Holland a quantity of carnations are grown for perfumery purposes and are extracted by volatile solvents.

Chemistry

To distillation 1000 kilos of carnation flowers yield 30 grams (0·003 per cent) of a pale green solid having an intense odour resembling that of the nine and ten carbon atom aldehydes. On extreme dilution this develops the true flower odour.

The chemistry of carnation flower oil has not received much attention, probably owing to its meagre yield. However, it has been studied by Glichitch who isolated from the distilled extract 31 per cent of stearophene which appeared to be identical with heptacosane. He removed traces of an aldehyde when the residual oil had an odour reminiscent of cinnamyl and citronellyl acetates. Treff and Wittrisch experimented upon clove pink blossoms grown at Gröba, in Saxony. They extracted 2·840 kg of flowers with petroleum ether and obtained a yield of 8·0 kg = 0·289 per cent of solid extract. This was treated with alcohol to remove inodorous matter, yielding 2·5 kg = 0·088 per cent of pure extract. This was finally steam distilled to yield 122·65 grams = 0·00432 per cent of

Odour classification

Top notes	Middle notes	Basic notes
1. Benzaldehyde	15. Heliotropin	65. Cinnamic alcohol
Benzyl acetate	16. Eugenol	79. Civet absolute
2. Rosewood	18. Orris absolute	88. Ethyl methyl phenyl glycidate
3. Terpineol	20. Clary sage	100. Amyl oxyiso-eugenol
4. Phenyl ethyl alcohol	21. Alpha ionone	Benzoin resin
6. Bergamot	22. Benzyl iso-eugenol	Decyl aldehyde
Iso-butyl phenylacetate	Clove	Iso-eugenol
8. Amyl salicylate	24. Geranyl acetate	Musk ketone
Benzyl salicylate	Rose otto	Pepper
Iso-butyl salicylate	Ylang	Peru balsam
Rhodinol	25. Methyl iso-eugenol	Phenyl acetic aldehyde
10. Methyl eugenol	30. Cardamon	Pimento
11. Carrot seed	31. Orange flower absolute	Vanillin
Nutmeg	41. Mace	
	42. Acetiso-eugenol	
	43. Eugenyl formate	
	Jasmin absolute	
	Rose absolute	
	Tuberose absolute	
	50. Neroly	
	Phenyl propyl aldehyde	

volatile oil. Upon chemical examination the following substances were identified:

	per cent
Eugenol	30
Phenylethyl alcohol	7
Benzyl benzoate	40
Benzyl salicylate	5
Methyl salicylate	1

Previous experiments in collaboration with Ritter gave the following yields:

	per cent
Solid extract	0·282
Pure extract	0·0926
Volatile oil	0·00498

On this occasion they surmised only the presence of eugenol.

Compounding notes

Carnation perfumes are based largely upon the isolates and synthetics derived from clove oil, of which eugenol is much favoured owing to its fragrant, peppery character. Salicylates are also indispensable constituents, and the vital natural ingredients are the rose alcohols, ylang and Peru balsam, reinforced with either orris absolute or carrot seed oil.

Two useful formulae follow:

Dianthus, no. 1005

30	Benzyl acetate
100	Phenyl ethyl alcohol
100	Iso-butyl salicylate
100	Rhodinol
10	Carrot seed
350	Eugenol
30	Benzyl iso-eugenol
80	Ylang
100	Iso-eugenol
100	Peru balsam
1000	

Oeillet, no. 1006

40	Terpineol
100	Amyl salicylate
30	Heliotropin
500	Eugenol
100	Clove
100	Iso-eugenol
10	Pepper
100	Peru balsam
20	Vanillin
1000	

Alcoholic Perfumes may be prepared as follows:

no. 1007

70	Dianthus
40	Oeillet
20	Rose absolute
10	Jasmin absolute
10	Benzoin resin
50	Musk tincture, 3 per cent
800	Alcohol
1000	

Cassie

History

Acacia Farnesiana is a small tree whose origin appears to be uncertain. It is said to be a native of San Domingo and became naturalised in Europe in the Farnesian gardens at Rome about 1656, but this date is probably incorrect since it is referred to by the author of a book published in Rome in 1625, entitled *Albini Hort. Farnesiana*. About 1764 Linnaeus, in *Hort. Upalensis*, described and named it *Mimosa Farnesiana*, which was afterwards, by Willdenow (1805), placed in his genus *Acacia*.

Varieties

There are several species of acacia which yield the typical cassie blossom odour. The more important of these are *Acacia Farnesiana*, Willd. (ancient cassie) and *A. Cavenia*, Hook. et Arn. (Roman cassie), both being cultivated in the south of France. The former is also cultivated in Syria, while it thrives in the Philippines, North and South America, India, Australia, Angola, Tunis, Egypt, and the West Indies; *A. Giraffoe* (the Camel tree) and *A. horrida* flourish in south-west Africa.

Cassie trees were first grown in the south of France at Cannes, then at Vallauris and Le Cannet. Owing to the increased popularity of Cannes as a winter resort large hotels and residences have been built on the fields once occupied by these trees, and consequently in recent years cultivation has moved further afield to Golfe-Juan, Mougins, and St. Laurent-du-Var.

The ancient and Roman cassie resemble one another, but the perfumes of the former is much finer, and thus the flowers command about twice the price of those of the latter. The difference in the appearance of the two will be readily appreciated by a close examination of the flowers. Both plants are affected by atmospheric conditions, but *A. Cavenia* has the advantage in being much less sensitive and requiring less attention during cultivation. Both require a sunny situation and protection from the cold winds; they thrive best on a light sandy granite soil. *A. Farnesiana* is grown from seed and *A. Cavenia* from cuttings, both being planted out during the month of March. The former must be pruned every year, in the spring, but only old and useless branches are removed in the latter. Nowadays *A. Farnesiana* is often grafted directly upon the hardier *A. Cavenia*. If the wood of these trees is destroyed by frost grafting again takes place on the shoots. In a couple of years the harvest becomes normal.

A. Farnesiana blossoms from the end of September to the beginning of February. Abundant crops are obtained if the temperature has been mild and moist. The trees are then trimmed.

A. Cavenia generally yields two crops each year, the first being collected at the same time as the above, and the second in the spring, sometimes as late as May.

The cassie blossoms (sometimes known locally as 'Pompons') are generally harvested by women, and great care is necessary during collection. The trees grow to a height of about 15 feet, and the blossoms are successive, some being ready for collection before others are scarcely formed. The flowers are gathered twice a week, in the daytime, and are conveyed to the works in the evening—the yield of blossoms per tree being from 2 to 5 kilos.

Odour

The species mentioned above bear golden globular flowers, whose perfume closely resembles a perfectly blended combination of orange blossom and violet with just a mere suggestion of cummin. As already stated those of *A. Cavenia* are not so fine.

Natural perfume

Cassie flowers are extracted by means of hot fats or by volatile solvents, and the absolute flower oil is prepared from both pomade and concrete. The latter process is now almost exclusively employed. About 250 kilos of flowers produce 1 kilo of concrete which in turn yields about 300 grams of absolute. The Middle East is now one of the most important sources: the concrete is made on the spot and shipped to Grasse for the preparation of the Absolute.

The flowers cultivated in Syria are not extracted by the above method, but the floral odours are absorbed by means of plates of spermaceti which are sent to France to be worked up into perfumes.

In practical perfumery cassie absolute enters largely into the preparation of numerous violet bouquets, and to these it imparts a delightful and peculiar fragrance unobtainable with any other product. In this connection it should be noted that when compounding synthetic violet ottos, the use of some natural violet absolute is recommended, but owing to its prohibitive cost something cheaper often has to be found to replace it. Cassie absolute is a good substitute and about 5 per cent may be used with satisfactory results.

Odour classification

Top notes	Middle notes	Basic notes
1. Benzyl acetate	15. Heliotropin	73. Cassie absolute
2. Linalol	16. Eugenol	77. Methyl naphthyl ketone
Methyl salicylate	18. Violet leaf absolute	80. Hydroxy citronellal
Rosewood	21. Alpha ionone	100. Costus
3. Cuminic aldehyde	Anisic aldehyde	Coumarin
Terpineol	Methyl anthranilate	Decyl aldehyde
4. Cummin	28. Cinnamyl butyrate	Iso-eugenol
Phenyl ethyl alcohol	31. Orange flower absolute	Orris resin
6. Bergamot	43. Jasmin absolute	Styrax resin
7. Geraniol, Java	50. Laurinic aldehyde	Tolu balsam
8. Iso-butyl salicylate	Phenyl propyl aldehyde	Vanillin
9. Geranyl butyrate		
11. Benzyl butyrate		
12. Methyl heptine carbonate		
Petitgrain, French		
13. Orris concrete		
14. Methyl ionone		
Mimosa absolute		

Chemistry

During the past thirty years practically no research work on this essential oil has been published. Prior to 1904 it had been studied by Schimmel & Co., von Soden, and Haarmann and Reiner, who identified the following constituents:

    Methyl salicylate   Anisaldehyde
    Benzyl alcohol   Benzaldehyde
    Farnesol   Cuminic aldehyde
    Linalol   Decyl aldehyde
    Geraniol

Traces of *p*-cresol exist and also two undefined ketones, one resembling menthone and the other having an odour recalling violet.

In cassie Romaine, Walbaum also found some eugenol and methyl eugenol.

Compounding Notes

The perfume of cassie is reminiscent of a blend of violet with orange blossom, and these flower oils may be liberally used in any compound of this type. But on smelling the flower the known constituents of eugenol, methyl salicylate and cuminic aldehyde will also be perceived, so that every good cassie should contain them. Moreover, there is a faint suggestion of limes blossom pervading the whole, and this subtle shading may be obtained by additions of hydroxy citronellal and terpineol.

Basic and compounded formulae follow:

Cassie, no. 1008

40	Linalol
20	Methyl salicylate
10	Cuminic aldehyde
150	Terpineol
10	Methyl heptine carbonate
20	Orris concrete
300	Methyl ionone
50	Mimosa absolute
100	Eugenol
70	Anisic aldehyde
10	Orange flower absolute
10	Cassie absolute
50	Methyl naphthyl ketone
150	Hydroxy citronellal
10	Decyl aldehyde, 10 per cent
1000	

Cassie, no. 1009

20	Methyl salicylate
10	Cuminic aldehyde
150	Terpineol
350	Violet compound
200	Dianthus compound
100	Orange blossom compound
30	Mimosa absolute
30	Cassie absolute
10	Jasmin absolute
100	Hydroxy citronellal
1000	

FLOWER PERFUMES: CHYPRE

Alcoholic perfumes may be prepared thus:

no. 1010

70	Cassie, no. 1008
70	Cassie, no. 1009
10	Violet leaf absolute, 10 per cent
20	Civet tincture, 3 per cent
30	Musk tincture, 3 per cent
800	Alcohol
1000	

Chypre

Strictly speaking, chypre perfumes are 'flowery' rather than 'flower' perfumes, and should not therefore be included in this series of monographs. Their importance today, however, commands a place, hence these notes.

History

The island of Cyprus is known to the French as Chypre, and the perfume appears to have originated in this island. There is no clear record as to dates, although it was in the twelfth century, at the time of the Crusades, that Richard I of England assumed the title of King of Cyprus, and, according to Piesse, eau de chypre was then introduced to Europe. Before this time the island was patronised largely by the elite of adjacent countries, such as Italy, Greece, Persia, and Egypt, and in view of the importance there of cistus plants (labdanum), it would not be surprising that a perfume should have been made with this still much-esteemed material. The word chypre was first given to a perfume in the fourteenth century, when 'oyselets de chypre' or 'cypre' were composed of labdanum, styrax, and calamus, made into a paste with tragacanth and then moulded in the form of a bird. They were very popular on the Continent, and were burned much in the manner as pastilles are today. The first notable modification in the basic composition of the perfume occurred towards the end of the fourteenth century when oakmoss was added. Dejean's *Traite des Odeurs*, 1777, gives two recipes for chypre. The first contains white, washed oakmoss in powder, with musk, ambergris, and civet. The second is more elaborate, containing oakmoss, orange

Odour classification

Top notes	Middle notes	Basic notes
1. Amyl acetate	15. Cinnamyl acetate	65. Cinnamic alcohol
2. Linalol	Heliotropin	73. Cassie absolute
Mandarin	16. Eugenol	79. Civet absolute
Methyl salicylate	18. Calamus	87. Octyl aldehyde
Rosewood	20. Clary sage	90. Estragon
3. Sassafras	21. Alpha ionone	91. Undecalactone
4. Phenyl ethyl alcohol	22. Clove	94. Angelica root
6. Bergamot	24. Cinnamon bark	100. Castoreum absolute
8. Amyl salicylate	Geranium Grasse	Coumarin
Cedarwood	Rose otto	Cypress
Iso-butyl salicylate	Ylang	Decyl aldehyde
Lemon	27. Dimethyl hydro-quinone	Iso-eugenol
9. Tagette		Labdanum resin
10. Linalyl acetate	43. Jasmin absolute	Methyl nonyl acetaldehyde
11. Sweet orange	Rose absolute	
13. Orris concrete		Musk ambrette
14. Basilic		Musk ketone
Methyl ionone		Oakmoss
		Patchouli
		Pimento
		Santal
		Styrax resin
		Undecylenic aldehyde
		Vanillin
		Vetivert

flowers, benzoin, storax, civet, almonds, cardamon, roses, clovewood, santal, and camphor.

Compounding notes

Chypre perfumes are generally classified as belonging to the extremely heavy, very clinging type, and curiously enough when copious quantities are worn by elegant ladies the peculiar fragrance is much appreciated by masculine taste. Oakmoss from all the usual sources, whether in the form of green or discoloured absolute or resin, is invariably employed in large percentage as one of the indispensable basic constituents, although some of the older established firms still use it as an alcoholic tincture of 10 per cent strength. Some years ago an enterprising Grasse manufacturer distilled the lichen over cedarwood oil, and offered the resulting product as an alternative in the belief that it would neither discolour nor darken with age after blending, but it was not a great success.

A reference to the odour classification will reveal all the basic notes that may be usefully blended with oakmoss to increase the depth of odour and long lasting qualities of the finished perfume. Those of vital importance are civet, estragon, castoreum, coumarin, labdanum, musk ambrette, patchouli, santal, vanillin, and vetivert. While blends of the Top and Middle Notes may be chosen according to the fancy of the artist, it is interesting to note that in recent years large quantities of sweet orange have been used to dominate the Top Note, and that undecalactone, methyl nonyl acetic and undecylenic aldehydes have also found great favour.

Examples of the older and newer types of perfume follow:

Chypre, no. 1011

60	Benzyl acetate
120	Phenyl ethyl alcohol
200	Bergamot
50	Ylang
120	Jasmin compound
100	Rose compound
10	Civet absolute
5	Angelica root
5	Castoreum absolute
120	Coumarin
30	Labdanum resin
20	Musk ambrette
10	Oakmoss
20	Patchouli
70	Santal
60	Vanillin
1000	

Modern Chypre, no. 1012

250	Bergamot
130	Sweet orange
5	Basilic
200	Methyl ionone
20	Rose otto
50	Jasmin absolute

10	Civet absolute
5	Estragon
3	Undecalactone
40	Coumarin
30	Dianthus compound
30	Labdanum resin
2	Methyl nonyl acetic aldehyde, 10 per cent
30	Musk ketone
30	Oakmoss
30	Patchouli
70	Santal
10	Undecylenic aldehyde, 10 per cent
5	Vanillin
50	Vetivert
1000	

A finished perfume may be prepared as follows:

no. 1013

130	Modern chypre, no. 1012
30	Ambergris tincture, 3 per cent
40	Musk tincture, 3 per cent
800	Alcohol
1000	

Cyclamen

History

This plant was known many years before Christ, but classical literature contains no references to its charming odour. There are, however, some interesting points in connection with its supposed medicinal value. Theophrastus calls attention to the properties of *Cyclamen graecum* as follows:[3] 'Of cyclamen the root is used for suppurating boils; also as pessary for women and mixed with honey for dressing wounds; the juice for purgings of the head for which purpose it is mixed with honey and poured in; it also conduces to drunkenness, if one is given a draught of wine in which it has been steeped. They say also the root is a good charm for inducing rapid delivery, and as a love potion when they have dug it up, they burn it, and then having steeped the ashes in wine, make little balls like those made of wine-lees which we use as Soap.' Pliny[4] discusses the properties of three different cyclamens,

[3] *Enquiry into Plants*, ix, 9, 3.
[4] Book XXV, chaps. 67, 68, and 69.

but only one of these is considered by commentators to belong to this genus of the N.O. Primulaceae, i.e. *Cyclamen hederoefolium*, the ivy-leaved cyclamen. The other two are now considered to be respectively the Italian honeysuckle, *Lonicera caprifolium*, and the small lily of the valley, *Convalleria bifolia*. Pliny calls the cyclamen 'tuber terrae', and while attributing to it many of the properties already mentioned by Theophrastus, states that when kept in the house noxious smells have no effect, and further that if a pregnant woman steps over the root she will be sure to miscarry!

The common name for cyclamen is Sowbread. It appears that in Italy and Sicily where the plant is fairly common, the fleshy rootstock (corm) is much appreciated by swine, hence the name *Pane porcino.*

Varieties

The cyclamen is easily recognised from its reflexed corolla. The plants are found growing wild in subalpine regions and in this country are much cultivated in greenhouses; hence the name 'cyclamen des Alpes'.

C. Europoeum is one of the best-known species with its reddish-purple flowers. The blooms appear from August to November, and a corm twenty years old is said to bear as many as one hundred flowers.

C. coum is native to southern Europe. Flowers rose or white appearing in February or March.

C. neapolitanum (*graecum*). Rose or white flowers with purplish tints at reflex point. Blossoms appear August to October.

C. repandum (*hederoefolium*). Ivy-leafed cyclamen. Rosered and white flowers appearing from March to May, common in Central Europe and northern coast of Africa. Corm not infrequently 10 to 12 inches in diameter, covered with cracked or scaled brownish rough rind.

Natural perfume

This is not an article of trade, and there appears to be no record of any attempts to extract it commercially. Experimentally, however, cyclamen received some attention from F. Elze, who macerated the flowers (species not stated) with liquid fat and obtained therefrom an extract by means of petroleum ether. After treatment with alcohol a dark coloured absolute was obtained in which this chemist identified, with certainty, nerol and farnesol. He states also that ketones, aldehydes, phenols, and esters were present. The perfumes of the flowers vary slightly. It is most powerful in *C. persicum*, and recalls a blend of lily-lilac and violet, with occasionally a suggestion of hyacinth.

Compounding notes

The perfume of cyclamen is too complex to suggest any one aromatic substance as the base, but a blend of hydroxy citronellal

Odour classification

Top notes	Middle notes	Basic notes
1. Benzyl acetate	15. Cinnamyl acetate	65. Cinnamic alcohol
2. Linalol	Guaiac wood	73. Cassie absolute
Phenyl ethyl acetate	Heliotropin	79. Civet absolute
3. Dimethyl benzyl carbinol	18. Orris absolute	80. Hydroxy citronellal
Terpineol	Phenoxyethyl iso-butyrate	88. Ethyl methyl phenyl glycidate
4. Citronellol	Violet leaf absolute	89. Cyclamen aldehyde
Phenyl ethyl alcohol	19. Phenyl ethyl iso-butyrate	91. Undecalactone
6. Bergamot	21. Anisic aldehyde	100. Ambergris
Citronellyl formate	Ionone beta	Amyl cinnamic aldehyde
Phenyl methyl carbinyl acetate	22. Geranyl formate	Benzoin resin
7. Geraniol, Java	24. Anisyl acetate	Coumarin
Methyl octine carbonate	Geranyl acetate	Decyl aldehyde
Violet absolute	Jonquille absolute	Guaiyl acetate
8. Amyl salicylate	Ylang	Musk ambrette
Benzyl salicylate	43. Jasmin absolute	Musk ketone
Nerol	Rose absolute	Oakmoss
Rhodinol	Tuberose absolute	Olibanum resin
10. Linalyl acetate	47. Anisic alcohol	Phenyl acetic acid
11. Linalyl cinnamate	50. Laurinic alcohol	Phenyl acetic aldehyde
14. Methyl ionone	Laurinic aldehyde	Santal
Mimosa absolute	Neroly	Styrax resin
	55. Nerolidol	Tonka resin
	60. Citral	Undecylic aldehyde
		Vetivert
		Vanillin

and methyl ionone in the ratio of 7 to 3 will simulate the lily-violet complex. This may be perfected by additions of terpineol and cinnamic alcohol to provide the lilac-hyacinth shading, while of course cyclamen aldehyde is an essential constituent and is enhanced in combination with laurinic aldehyde. The slightly earthy background noticeable in some flowers may be obtained with traces of oakmoss.

Two formulae based on these ideas follow:

Cyclamen, no. 1014

60	Terpineol
50	Bergamot
1	Methyl octine carbonate
100	Nerol
150	Methyl ionone
100	Heliotropin
30	Jasmin absolute
20	Rose absolute
1	Laurinic aldehyde
100	Cinnamic alcohol
350	Hydroxy citronellal
5	Cyclamen aldehyde
30	Musk ketone
2	Oakmoss
1	Phenyl acetic aldehyde
1000	

Cyclamen, no. 1015

180	Lilac compound
50	Bergamot
100	Nerol
180	Violet compound
20	Jasmin absolute
10	Rose absolute
1	Laurinic aldehyde
420	Muguet compound
7	Cyclamen aldehyde
10	Musk ketone
20	Jacinthe compound
2	Oakmoss
1000	

Alcoholic Perfumes may be prepared as follows:

Cyclamen, no. 1016

140	Cyclamen compound, no. 1014
3	Rose otto
5	Jasmin absolute
2	Cassie absolute
20	Ambergris tincture, 3 per cent
30	Musk tincture, 3 per cent
800	Alcohol
1000	

Fern

History

Ferns have been known from time immemorial and are mentioned fairly frequently in classical literature. A number of them are described and compared by Theophrastus, but he makes no mention of any pleasant smelling varieties. He draws attention to both white and black maiden-hair—called wet-proof—because it does not catch the dew nor get wet when watered. It was found growing in damp places, and when pounded up and mixed with olive oil was used by the Greeks to prevent hair falling out. Other ferns mentioned by this author are—bracken, *Pteris aquilina; nephrodium felix-mas* and heart's tongue, *Scolopendrium vulgare*. Pliny mentions two kinds which are equally destitute of blossom and of seed. One of these, he says, has a root with a not unpleasant smell. He describes fully the anthelmintic properties of the male fern. The common bracken has in the past been much esteemed. The cut and dried fronds have been used for domestic purposes. The underground stems contain a quantity of mucilage and starch which in some parts of Europe and northern countries are prepared by washing and pounding and are mixed with meal to make bread in times of scarcity. At one time a distinct species, *Pteris esculenta*, formed an important item of food to the natives of the Pacific Islands. The absence of visible flowers and seeds on ferns has attached much superstition to these plants. In Shakespeare's time they were spoken of as 'uncanny and evil', and it was considered that those who possessed fern seed could make themselves invisible at pleasure. A more practical notion of the supposed power of ferns is, that the burning of it brings down rain, of which the following is a curious illustration. In a volume containing a miscellaneous collection by Dr Richard Pocock, in the British Museum, is the copy of a letter written by Philip Herbert, third Earl of Pembroke, Lord Chamberlain to the sheriff of Staffordshire. It is as follows: 'Sir, His Majesty, taking notice that the burning of Ferne doth draw down rain, and being desirous that the country and himself may enjoy fair weather as long as he remains in these parts, His Majesty has commanded me to write to you to cause all burning of Ferne to be forborne until His Majesty be past the country. Wherein, not doubting but the consideration of their own interest as well as of His Majesty's will invite the country to a ready observance of this His Majesty's commands. I rest, your very loving friend, Pembroke and Montgomery.'

Varieties

Ferns, having no true flowers, belong to the division of plants known as Cryptograms, the natural order of this particular branch being *Felicineae*, of which nearly 3000 species have been named and described. They are widely distributed over the globe and vary in size and form from a small plant to a lofty tree of 50 feet in height terminated by a crown of finely-cut leaves, termed fronds, varying in length, and sometimes in the larger species exceeding 15 feet. Reproduction is by means of spores, generated on the under side of the fronds. Commercially the most important species is male fern, *Dryopteris Felixmas*, abundant in Britain and one of the commonest of our indigenous ferns. The rhizome is collected in the autumn, and the fronds and roots removed. It is dried and subsequently extracted with volatile solvents. This fern is common also in Germany and the Harz Mountains. The extract is, as is well known, used medicinally and to some extent in Fougère perfumes. Other interesting ferns are:

Adiantum amabile. Scented maiden-hair—an elegant stove fern from Brazil. Young specimens have a pleasant fragrance.
Asplenium fragrans, growing wild in North America, fronds dried and used as bedding.
A. Onopteris and *A. Obovatum*, common to the shores of the Mediterranean and particularly the Esterel.
Osmunda regalis, common to bogs, woods, and wet meadows in Europe.

Other species found in sheltered parts of the south of France are *Polypodium vulgare* and *Scolopendrium officinale.*

Chemistry

The natural perfume is an article of commerce, though to perfumers not very common since they generally prefer to base Fougère perfumes upon oakmoss and patchouli. The male fern rhizome contains about 5 per cent of *Filmarone,* a yellow, amorphous substance of acid nature, to which the vermifuge properties of the drug appear to be attributed. The volatile solvent extract contains in addition aspidinol, flavaspidic acid, flavaspidinol, albaspidin, and filicitannic acid. It should be of a greenish rather than a brownish colour.

Compounding Notes

Fougère perfumes might almost be classified as deep-noted, sophisticated lavenders. They are much appreciated by masculine taste, and especially so in the form of toilet waters. Lavender oil, therefore, dominates the Top Note in all perfumes of this type,

Odour classification

Top notes	Middle notes	Basic notes
1. Acetophenone	15. Acet anisol	73. Cassie absolute
Benzyl acetate	Heliotropin	79. Civet absolute
Methyl benzoate	20. Clary sage	90. Estragon
2. Linalol	21. Anisic aldehyde	100. Benzophenone
Methyl salicylate	Ionone alpha	Castoreum
Rosewood	Myrrh resin	absolute
3. Methyl	Rosemary	Coumarin
acetophenone	24. Geranium Grasse	Musk ambrette
Sassafras	Rose otto	Musk ketone
4. Eucalyptus	Ylang	Oakmoss resin
Lavender	27. Dimethyl hydro-	Patchouli
6. Bergamot	quinone	Santal
7. Benzyl propionate	43. Jasmin absolute	Tonka resin
8. Amyl salicylate	Rose absolute	Undecylenic
Cedarwood	Tuberose absolute	aldehyde
Iso-butyl benzoate	45. Bornyl acetate	Vanillin
Iso-butyl salicylate		Vetivert
10. Linalyl acetate		
12. Petitgrain, French		

FLOWER PERFUMES: FERN

and is improved by blending with bergamot and rosewood. Jasmin, rose and cassie absolutes provide the floral shading, but it is the Basic Note that is of paramount importance. Thus, coumarin, oakmoss, patchouli, santal and vetivert are constituents of almost all formulae, and in some of them traces of undecylenic aldehyde not only reinforce the note but also afford a peculiar and inimitable quality.

Formulae for a basic compound, perfume and toilet water follow:

Fougère, no. 1017

5	Acetophenone
100	Benzyl acetate
100	Rosewood
200	Lavender
300	Bergamot
20	Amyl salicylate
10	Clary sage
10	Anisic aldehyde
5	Rose otto
30	Jasmin absolute
3	Civet absolute
70	Coumarin
30	Musk ambrette
20	Oakmoss
30	Patchouli
40	Santal
2	Undecylenic aldehyde
5	Vanillin
20	Vetivert
1000	

Fougère Perfume, no. 1018

140	Compound, no. 1017
10	Tonka resin
10	Ambergris tincture, 3 per cent
40	Musk tincture, 3 per cent
800	Alcohol
1000	

Fougère Toilet Water, no. 1019

30	Compound no. 1017
20	Musk tincture, 3 per cent
100	Distilled water
850	Alcohol
1000	

Gardenia

Botany

The gardenia, as generally known in this country, is the flower of *Gardenia Florida L.,* a shrub belonging to the N.O. Rubiaceae, and native to tropical Asia and South Africa. The genus was named in honour of Dr Garden, of Charleston, California, a correspondent of Linnaeus, and consists of numerous species, all of which bear beautiful and highly fragrant flowers. Amongst these may be mentioned:

G. Florida, known also as Cape jessamine, has double white flowers, fragrance reminds of jasmin, with trace phenyl methyl carbinyl acetate. Known in China as pak-sema-hwa and used for scenting tea. Berry orange coloured, and size of pigeon's egg. Pulp used in Far East for dyeing yellow.

G. Radicans, native of Japan. Dwarf free blooming variety.

G. Calijculata, native of mountains in India, large white flowers. In Bengal known as gundhuraja.

G. Costata, known also as *G. Coronaria,* tree of 25 feet high, large salver-shaped flowers, tube 3 inches long, border 4 inches diameter. First grown Botanic Garden of Calcutta.

G. Grandiflora, large white flowers. Native Cochin China on banks of rivers.

G. Tomentosa, native of Java.

G. Devoniana, native of Sierra Leone.

G. Citriodora, odour recalls syringa and orange blossom.

G. Thunbergia like *G. Floribunda,* native of south-west Africa.

G. Gummifera and *G. lucida,* native of East Indies. Wounding of bark exudes a fragrant resin not unlike elemi.

The species of the genus *Randia* are not unlike gardenias and many of them are highly fragrant.

Some years ago Guido Mariotti, an Italian professor, experimented with the hybridisation of gardenias, and succeeded in producing thirteen coloured and scented specimens *shaped like roses.*

Odour

The absolute of gardenia is not a Grasse commercial product, but through the courtesy of Mr Louis Amic of Roure-Bertrand Fils, the author received a small sample especially prepared from the flowers by means of petroleum ether. This absolute was yellowish-brown in colour, semi-liquid in consistence, and had an odour recalling that of a mixture of the absolutes of jasmin, orange blossom, and tuberose. An odour suggesting phenyl methyl carbinyl acetate could just be detected.

Natural Perfume

This is now regularly produced in Reunion, and was first experimented with by Charles Garnier in 1912. The variety of flower cultivated is not stated, but the gardenia fields are situated at a high altitude where moisture is favourable to growth and cyclones are ineffective. There are roughly 7000 plants per hectare. Cuttings are placed over one metre apart and a well-developed bush yields from 100 to 260 grams of blossoms per annum. The harvest is collected in November and December. Extraction with volatile solvents yields the commercial product which is expensive. One kilo of concrete is obtained from 3000 to 4000 kilos of flowers, and this again yields to alcohol about 500 grams of absolute. This product even after careful purification by freezing, deposits yellowish-white crystals on the inside of the container. On standing, a yellowish semi-waxy mass separates and floats on the surface. The odour bears a close resemblance to that of the natural flower.

Chemistry

This was investigated by E. Parone, who treated 250 kilos of fresh flowers by maceration in liquid vaseline oil and shook out the essential oil with absolute alcohol, obtaining a yield of 176 grams having a yellowish colour. He identified the following constituents: benzyl acetate, styrolyl acetate, linalol, linalyl acetate, terpineol and methyl anthranilate, together with traces of benzoic acid as ester. Although the chief constituent was benzyl acetate, the characteristic odour was due to styrolyl acetate.

Compounding Notes

The perfume of gardenia is reminiscent of that of tuberose whose heavy character is diminished by the lighter fragrances of jasmin and orange blossom, and in which suggestions of the powerful synthetic, phenyl methyl carbinyl acetate, are apparent. Thus, blends of these flower compounds, crowned by gardenol, are an

Odour classification

Top notes	Middle notes	Basic notes
1. Benzyl acetate	15. Cinnamyl acetate	65. Cinnamic alcohol
Methyl benzoate	Heliotropin	73. Cassie absolute
2. Benzyl formate	16. Eugenol	77. Methyl naphthyl ketone
Ethyl benzoate	18. Orris absolute	
Linalol	20. Clary sage	79. Civet absolute
Methyl salicylate	21. Anisic aldehyde	80. Hydroxy citronellal
Octyl acetate	Indole	
Rosewood	Ionone alpha	85. Phenyl acetaldehyde
Phenyl ethyl acetate	Methyl anthranilate	
3. Coriander	Myrrh resin	Dimethyl acetal
Cyclohexanyl butyrate	22. Benzyl iso-eugenol	87. Octyl aldehyde
	Clove	88. Ethyl methyl phenyl glycidate
Decyl formate	23. Broom absolute	
Terpineol	24. Geranyl acetate	89. Cyclamen aldehyde
4. Methyl butyrate	Jonquille absolute	
Nonyl aldehyde	Ylang	91. Undecalactone
Phenyl ethyl alcohol	30. Ambrette seed	100. Acet eugenol
	31. Orange flower absolute	Amyl cinnamic aldehyde
6. Bergamot		
Phenyl methyl carbinyl acetate	42. Acetyl iso-eugenol	Benzoin resin
	43. Jasmin absolute	Castoreum absolute
7. Benzyl iso-butyrate	Rose absolute	
Benzyl propionate	Tuberose absolute	Costus
Methyl octine carbonate	50. Laurinic alcohol	Coumarin
	Laurinic aldehyde	Decyl aldehyde
8. Amyl salicylate	Phenyl propyl aldehyde	Gamma nonyl lactone
Benzyl salicylate		
Geraniol palmarosa	Neroly	Iso-eugenol
Nerol		Labdanum resin
Rhodinol		Methyl nonyl acet aldehyde
11. Linalyl cinnamate		
Sweet orange		Musk ambrette
12. Methyl heptine carbonate		Musk ketone
		Patchouli
Petitgrain, French		Peru balsam
13. Phenyl propyl alcohol		Phenyl acetic aldehyde
14. Methyl ionone		Santal
Mimosa absolute		Tolu balsam
Phenyl propyl acetate		Vanillin
Phenyl propyl iso-butyrate		Vetivert

FLOWER PERFUMES: GARDENIA

easy and direct means of preparing this perfume. But since many experimenters prefer to work with basic raw materials, the following constituents would seem to be vital:

Top Notes. Benzyl acetate, linalol, terpineol, phenyl ethyl alcohol, bergamot and phenyl methyl carbinyl acetate.

Middle Notes. Methyl anthranilate, ylang, jasmin and tuberose absolutes and neroly.

Basic Notes. Hydroxy citronellal, amyl cinnamic aldehyde, decyl aldehyde and gamma nonyl lactone.

Formulae illustrating each type follow:

Gardenia, no. 1020

30	Nerol
20	Linalol
40	Bergamot
20	Phenyl methyl carbinyl acetate
50	Ylang
200	Orange blossom compound
200	Jasmin compound
400	Tuberose compound
30	Dianthus compound
10	Methyl nonyl acetic aldehyde, 10 per cent
1000	

Gardenia, no. 1021

150	Benzyl acetate
100	Linalol
5	Methyl salicylate
60	Terpineol
100	Phenyl ethyl alcohol
80	Bergamot
30	Phenyl methyl carbinyl acetate
10	Methyl anthranilate
70	Ylang
30	Jasmin absolute
50	Tuberose absolute from pomade
20	Neroly
30	Methyl naphthyl ketone
200	Hydroxy citronellal
5	Ethyl methyl phenyl glycidate
30	Amyl cinnamic aldehyde

10	Decyl aldehyde, 10 per cent
15	Gamma nonyl lactone
5	Phenyl acetic aldehyde
1000	

A finished perfume may be prepared from the foregoing as follows:

Gardenia, no. 1022

50	Gardenia compound, no. 1020
100	Gardenia compound, no. 1021
10	Civet absolute, 10 per cent
40	Musk tincture, 3 per cent
800	Alcohol
1000	

Hawthorn

History

May blossom, *Crataegus oxycantha*, of the N.O. Rosaceae, has been known since the earliest times, and was mentioned by Theophrastus as growing on Mount Ida. It was regarded by the Greeks as a tree of fortune, while the Romans used it as a symbol of marriage. Its flowering branches are said to have been borne aloft at their weddings, and the newly-wedded pairs were even lighted to their nuptial chambers with torches of its wood. May Day has for years been one of our rural festivals, and many a pretty scene must have been enacted in London during the time of Robin Hood and Friar Tuck, for we are told that the doors were decked with hawthorn and that fantastic dancers performed their antics round the maypole. There was so much enthusiasm at this period that young men and maidens went to the fields at sunrise to wash their faces in the dew and gather the treasured may.

The hawthorn has a traditional connection with the royal House of Tudor. When Richard III was slain at Bosworth Field, a small crown of gold which he wore as a crest on his helmet was found by a soldier on a bush of hawthorn. It was brought to the newly-made King Henry VII, the first royal Tudor, on whose head it was placed when the army saluted him as their sovereign. It is said that in memory of this event the House of Tudor assumed the device of a crown in a bush of fruited hawthorn.

Varieties

There are about forty species of this attractive genus of shrubby trees, which are widely spread over both hemispheres. The height attained by them is about 20 feet, and they live to a great age. The main floral distinctions are colour and single or double flowers, while there is also a variation in the tint of the fruits. The common white-may of our English hedgerows is the most fragrant, and is not approached by the variety with scarlet flowers, which occasionally emit a disagreeable odour. The early spring variety, sometimes called the glastonbury thorn, is known to have flowered just after Christmas, when the atmospheric conditions have been favourable. Another variety, *C. crusgalli*, known as the cockspur thorn and coming from North America, attains a height of about 10 feet only and is remarkable for its peculiar growth. This is particularly noticeable in *C. pyracanthifolia* where the branches spread out like a table. The older the tree, the more pronounced this feature becomes. Another species, *C. aronia*, is common in what was known as Palestine, and particularly near Jerusalem, where its red, fleshy fruits are collected for preserves.

Odour

The exquisite fragrance of the hawthorn is so well known that it requires very little comment. The odour is spicy and recalls that of almonds. The natural perfume is not an article of commerce.

The odour of hawthorn blossom is almost duplicated in the following:

Erica arborea
Pyrus communis
Sorbus acuparia
Viburnam tinus

In all these the presence of anisic aldehyde would seem to be indicated, but so far no chemical proof has been deduced.

Compounding notes

The fragrance of May blossom is well represented by anisic aldehyde, but the note is modified slightly with methyl acetophenone and strengthened with coumarin. Thus, these three synthetics are indispensable constituents of all compounds of this

Odour classification

Top notes	Middle notes	Basic notes
1. Acetophenone	15. Acet anisol	65. Cinnamic alcohol
Benzaldehyde	Heliotropin	73. Cassie absolute
2. Linalol	21. Anisic aldehyde	79. Civet absolute
3. Methyl acetophenone	Ionone alpha	80. Hydroxy citronellal
Terpineol	24. Anisyl acetate	
4. Citronellol	Jonquille absolute	100. Benzoin resin
Nonyl aldehyde	Rose otto	Coumarin
6. Bergamot	Ylang	Decyl aldehyde
Iso-butyl phenylacetate	27. Dimethyl hydroquinone	Musk ketone
7. Geraniol, Java	31. Orange flower absolute	Orris resin
Nonyl alcohol	43. Jasmin absolute	Peru balsam
8. Nerol	Rose absolute	Phenyl acetic aldehyde
Amyl salicylate	47. Anisic alcohol	Styrax resin
12. Petitgrain, French	60. Rhodinyl formate	Vanillin
		Vetivert

type, but their rather crude odour requires sweetening with ample proportions of rose alcohols or esters, together with a first quality lilac containing natural jasmin.

An example follows:

May blossom, no. 1023

10	Methyl acetophenone
40	Linalol
300	Anisic aldehyde
70	Ionone
80	Rhodinyl formate
450	Lilac compound
50	Coumarin
1000	

A finished perfume may be prepared as follows:

May blossom, no. 1024

130	Compound, as above
10	Bergamot
2	Orange flower absolute
3	Jasmin absolute
4	Rose absolute
1	Civet absolute

20	Ambergris tincture, 3 per cent
30	Musk tincture, 3 per cent
800	Alcohol
1000	

Heliotrope

History

Reference is made to this plant by many ancient writers among whom we may mention Theophrastus and Ovid. We are told by the former that it bloomed for a long period, and that the time of its flowering depended upon the heavenly bodies. The plant was supposed to have owed its existence to the death of Clytie, who pined away in hopeless love of the god Phoebus (Apollo), and the latter (Ovid) alludes to this in the following lines (*Metamorphoses*, Book IV, 255-270):

'But Clytie, though love could excuse her grief, and grief her tattling, was sought no more by the great light-giver (Apollo), nor did he find aught to love in her.

'For this cause she pined away, her love turned to madness. Unable to endure her sister nymphs, beneath the open sky, by night and day, she sat upon the ground, naked, bareheaded, unkempt.

'For nine whole days she sat, tasting neither drink nor food, her hunger fed by naught save her falling tears, and moved not from the ground.

'Only she gazed on the face of her god as he went his way, and turned her face towards him.

'They say that her limbs grew fast to the soil, and her deathly pallor changed in part to a bloodless plant; but in part 'twas red, and a flower, much like a violet, came where her face had been. Still, though roots hold her fast, she turns ever towards the sun and, though changed herself, preserves her love unchanged.'

The plant, whose flowers emit the most delightful fragrance, is *Heliotropium Peruvianum*, belonging to the N.O. Boraginaceae. It is a native of Peru, and was introduced into Europe about 1757.

Varieties

The genus *Heliotropium* contains about ninety species, of which the majority are botanically identified. They are distributed over the tropical and subtropical regions of both hemispheres and many are found growing in Europe. In England the plant never attains large proportions, but in southern France it grows to a bush 8 or 10 feet high. The species mentioned above is a favourite for greenhouse cultivation in this country, and the small clusters of its lavender-coloured flowers emit a delightful fragrance, which is somewhat almondy and inclined to be heavy. It is known by a variety of names, including heliotrope, cherry-pie, and Peruvian turnsole, the latter from the ancient fancy that it turned with the sun (see Ovid's reference to this above). This plant should not be confused with *Chrozophora tinctoria*, an annual of the N.O. Euphorbiaceae and native of southern Europe, where it is cultivated for the sake of a dye called *Turnsole*. The varieties of heliotrope include the following:

The Queen, flowers almost white.
Bouquet de Violettes, blooms of different shades of blue.
Bouquet Parfumé, flowers lilac-blue.

Winter heliotrope, *Tussilago fragrans*, is a native of southern Europe and blossoms early in December. The purple flowers are fragrant and resemble those of *Heliotropium Peruvianum*.

Heliotrope of the lowlands, *Lantana Camara*, L. (N.O. Verbenaceae), is indigenous to tropical South America and has been acclimatised in southern India. In Bombay it is known as 'Ghaneri', and the flowers have a sage-like odour.

Odour

The perfume of the heliotrope, as we have said, is 'almondy', and it is probably on this account that the plant received the name cherry-pie. The natural perfume is seldom met with in the form of an absolute.

Chemistry

Very little is known of the chemistry of the heliotrope perfume, but in 1876 Haarmann and Tiemann while examining the plant came to the conclusion that its odour was due principally to heliotropin and vanillin. The majority of heliotrope perfumes are, of course, based on these two synthetics and not on the natural heliotrope flower perfume.

FLOWER PERFUMES: HELIOTROPE

Odour classification

Top notes	Middle notes	Basic notes
1. Benzaldehyde	15. Heliotropin	65. Cinnamic alcohol
Benzyl acetate	21. Anisic aldehyde	79. Civet absolute
2. Benzyl formate	Ionone alpha	91. Undecalactone
3. Benzyl cinnamate	24. Anisyl acetate	100. Amyl cinnamic
4. Nonyl aldehyde	Rose otto	aldehyde
Phenyl ethyl alcohol	Ylang	Benzoin resin
6. Bergamot	31. Orange flower	Coumarin
7. Geraniol, Java	absolute	Methyl nonyl
11. Sweet orange	43. Jasmin absolute	acetic aldehyde
13. Orris concrete	Rose absolute	Musk ketone
	Tuberose absolute	Musk xylene
	50. Neroly	Patchouli
		Peru balsam
		Phenyl acetic
		aldehyde
		Santal
		Tonka absolute
		Vanillin
		Vetivert

Compounding notes

The richly balsamic fragrance of heliotrope is almost matched by heliotropin, but not quite; as developers are required to successfully create an acceptable compound. Vanillin is admirably suited for this purpose, together with either benzaldehyde or anisic aldehyde, but strangely enough the blend still needs an orange note to impart life to the whole. This may be chosen from sweet orange oil, orange flower absolute or methyl nonyl acetic aldehyde, and a further improvement is obtained by the use of a small percentage of rose otto.

Examples of compound and perfume follow:

Heliotrope Compound, no. 1025

20	Geraniol, Java
20	Sweet orange
250	Heliotropin
20	Anisic aldehyde
10	Rose otto
10	Musk xylene
20	Peru balsam

150	Vanillin
500	Diethyl phthalate
1000	

Heliotrope Perfume, no. 1026

140	Compound, no. 1025
1	Orris absolute, 10 per cent
5	Jasmin absolute
3	Rose absolute
1	Methyl nonyl acetic aldehyde, 10 per cent
20	Ambergris tincture, 3 per cent
30	Musk tincture, 3 per cent
800	Alcohol
1000	

Honeysuckle

History

Among the many unexplained superstitions of by-gone days is the one in connection with the woodbine, for it is stated by ancient writers that the herb-diggers thought they would suffer bodily harm if they did not dig up the roots of this plant before sunrise. Pliny refers to the honeysuckle,[5] saying that the *Clymenus*, as it was then called, was named after the son of Caeneus, King of Arcadia. It had leaves like those of ivy, numerous branches and a hollow stem. The smell of it was powerful and the seed like that of ivy; it grew in wild and mountainous localities. Pliny goes on to say that when taken in drink it cures certain maladies in the male sex, but in doing so it neutralises the generative powers!

The honeysuckle, *Lonicera*, is a twining shrub of the N.O. Caprifoliaceae, native of the temperate regions of the northern hemisphere and often found growing wild in the hedgerows of England. It is a favourite for planting against our countryside houses and frequently forms a graceful covering for arbours, porches, etc.

A peculiar feature about this plant is that of turning from east to west. The pressure exerted in the course of its twining growth is often sufficient to leave a well-marked indentation on the young trees supporting it.

[5] Book XXX, chap. 33.

Varieties

The genus consists of about eighty species, of which the following may be noted:

L. periclymenum, large creamy flowers, blooming early in the year—perfume most noticeable towards evening.

L. serotinum, reddish flowers, late in summer and autumn.

L. caprifolium, goats are said to have a predilection for its leaves—prolific in the south of France.

L. fragrantissima, bears white fragrant flowers about February.

L. brachypoda, yellow flowers, blooming from May to October. Known as Chinese honeysuckle.

L. sempervirens is a cultivated ornamental climber called trumpet honeysuckle.

L. gigantea or *L. santi* is the most common highly odorous variety found growing on the Riviera. The flowers are at first white and ultimately turn a deep yellow.

There are two other 'honeysuckles' which should not be confused with the above, namely:

Hedysarum coronarium, known as the French honeysuckle and largely grown on the Continent for feeding stock. It belongs to the N.O. Leguminosae.

Banksia australis is an Australian species of the N.O. Proteaceae, and named honeysuckle on account of the unusual amount of sweet liquid, like honey, contained in the flowers, which are sucked by the natives. It is stated to be so abundant in *B. ericifolia* and *B. Cunninghamii* that when in flower the ground underneath large cultivated plants is in a complete state of puddle.

Odour

The perfume of the woodbine is delightfully sweet, and one writer goes so far as to say that the fragrance of the violet does not compare favourably with it. Poets have always shown a preference for it, and in the lines of Mott we read:

> By rustic seat or garden bower
> There's not a leaf or shrub or flower
> Blossoms on bush so sweet as thee,
> Lowly, but fragrant honey tree.

The honeysuckle fragrance is particularly heavy towards evening and recalls tuberose, jasmin, and orange blossom in perfect blend.

This odour has a great attraction for nocturnal moths and crepuscular insects, and incidentally is the means of assuring their fecundation. The nectaria are placed at the base of the tubular corolla and cannot be reached by bees and diurnal butterflies who have a short proboscis. In consequence, their nectar is chiefly reserved for the sphinxes, particularly the *Sphinx ligustri*, the trunk of which is 4 to 5 cm in length.

Natural perfume

This is not yet an important commercial product, although the plant is specially cultivated for the production and utilisation of the flower in the south of France where a small quantity of concrete is prepared. In the Grasse district the blossoms appear during May and June. Antoine Chiris treated flowers of *Lonicera gigantea*, gathered about the middle of June, with petroleum ether. They yielded 3·3 per cent of a dark green, brittle, concrete essence, having an odour less fragrant than the flower. This concrete gave 23·8 per cent of an olive-green absolute, having a syrupy consistency, by the usual method. Steam distillation of this absolute resulted in 9 per cent of essential oil (0·7 per cent from the flowers). This was a limpid yellowish liquid having a penetrating odour, unpleasant at first and later becoming more fragrant

Top notes	*Middle notes*	*Basic notes*
1. Benzyl acetate	15. Cinnamyl acetate	65. Cinnamic alcohol
2. Linalol	Heliotropin	77. Methyl naphthyl ketone
3. Decyl formate Terpineol	21. Ionone alpha Methyl anthranilate	80. Hydroxy citronellal
4. Phenyl ethyl alcohol	24. Ylang	100. Amyl cinnamic aldehyde
6. Bergamot	43. Jasmin absolute Rose absolute	Coumarin
7. Geraniol, Java Methyl octine carbonate	50. Neroly Phenyl propyl aldehyde	Decyl aldehyde Musk ketone
8. Rhodinol Nerol		Olibanum resin Phenyl acetic acid
13. Para cresyl phenylacetate		Styrax resin Tolu balsam
14. Mastic Reseda absolute		Undecylic aldehyde Vanillin

but lacking the sweet character of the flowers. On examination it contained neither aldehydes, ketones nor nitrogen.

In the course of a visit to Grasse, the author discussed the problem of honeysuckle with Mr. Hubert Schleinger of Bertrand Frères, who very kindly undertook to further investigate this question. He purchased 205 kilos of flowers and treated them with petroleum ether by the volatile solvent extraction process, obtaining 590 grams of a concrete which yielded 320 grams of absolute flower oil. This had a pleasant odour incompletely reminiscent of the natural flower.

Compounding notes

For some strange reason honeysuckle perfumes have never achieved popularity, and in consequence there is little demand for the compound, save as a flowering agent in place of either lilac or muguet which are widely used for this purpose in fancy bouquets. The penetrating honey-like fragrance is reminiscent of a blend of orange blossom with rose, crowned with undecylic aldehyde and with a slightly balsamic background. Such a compound might be created on the following lines:

Honeysuckle, no. 1027

20	Benzyl acetate
200	Linalol
100	Nerol
20	Cinnamyl acetate
100	Heliotropin
20	Methyl anthranilate
30	Jasmin absolute
20	Neroly
300	Cinnamic alcohol
20	Methyl naphthyl ketone
150	Hydroxy citronellal
10	Undecylic aldehyde
10	Phenyl acetic acid
1000	

A finished perfume may be prepared from the above as follows:

Honeysuckle perfume, no. 1028

140	Compound, no. 1027
5	Orange flower absolute
4	Rose absolute

Honeysuckle perfume, no. 1028 (*continued*)

1	Civet absolute
50	Musk tincture, 3 per cent
800	Alcohol
1000	

Hyacinth

History

The hyacinth, *Hyacinthus orientalis*, belongs to the N.O. Liliaceae, and is a native of Syria and other parts of western Asia. It is believed to be the plant so frequently referred to in ancient classical literature. Homer describes it as a fragrant, bell-shaped flower, and says it took the foremost place on the mass of fragrant blossoms that formed the couch of Jupiter and Juno. From what Pliny says in Book XXI, chapters 39 and 97, and in Book XXV, chapter 80, it is fairly clear that under the name of hyacinth he has confused the characteristics of two different plants. The hyacinth, too, of Dioscorides, Book III, chapter 5, is a different plant, being, probably, *Hyacinthus comosus*. The Greek virgins all wore crowns of hyacinths when assisting at the weddings of their friends, and Ovid (Book X) attributes its origin to Hyacinthus, a beautiful Spartan youth, son of Amyclas, King of Amyclae, who was beloved by Apollo, and accidentally killed by the god in a game of quoits. From his blood a flower sprang up whose petals bore the marks of Apollo's grief. An annual solemnity, called Hyacinthia, was established in Laconia in honour of Hyacinthus. It lasted three days, during which the people, to show their grief for the loss of their beloved prince, ate no bread but fed upon sweetmeats, and abstained from adorning their hair with garlands as on ordinary occasions. The following day was spent on feasting, hence, perhaps, the floral meaning 'Play' often attributed to this flower. The poets of the Orient are fond of using the hyacinth for purposes of simile; Hafiz, for instance, compares his mistress's hair with the flower. Hyacinth locks are expressive of graceful tresses, because the petals of the flower turn up at the points.

The hyacinth was introduced to Britain during the sixteenth century, and was at that time a single-flowered species, but during the seventeenth century double-flowered ones began to appear. Four varieties were then recognised, the single and double blue, the purple and the violet. It is extensively cultivated in Holland, particularly at Haarlem, where there are large areas devoted entirely to the growth of this and other bulbous plants. The explanation of this extraordinary development in horticulture appears to be due to the peculiar adaptability of the soil and climate of Holland. When these huge areas are covered with blooms they present a wonderful sight, and the atmosphere for miles around becomes impregnated with their exotic fragrance.

Varieties

The genus *Hyacinthus* comprises about thirty species, the varieties of which through cultivation have become numerous. At least 4000 varieties have been reported, but commercially not more than 5 per cent are exploited. *H. orientalis* is the principal cultivated species, while *H. non-scriptus*, the blue-bell, and *Muscari racemosum*, the grape-hyacinth, are often found growing wild in this country. On the shores of the Mediterranean and in the Pyrenees, there is a small flowering species, *H. amethystinus,* having pretty bright blue flowers.

In the south of France the wild blue hyacinth is extracted by the Grasse works. Some manufacturers in Holland have taken up the production of absolute, and, as they say, it is worthy of much wider favour.

Odour

The odour of hyacinths is heavy but has been described as ethereal. This probably may be accounted for by the fact that the early flowering plants have a finer odour than the later flowering kinds, and also that just when the blossoms appear the perfume is much more delicate. If a plant is placed in a room or any confined space the odour becomes overpowering.

Natural perfume

This is extracted on a small scale by means of volatile solvents, and both the wild and cultivated flowers are used. The lighter coloured varieties yield the finest perfume and, commercially, the single blossoms are found to give the best results. Very small quantities are available, since growers have found it more remunerative to sell the flowers for decorative purposes or, as in most cases, to sell the bulbs. The production of the natural perfume, however, is not a commercial success.

Chemistry

The chemistry of the hyacinth perfume is not yet completely understood. Some years ago, however, an essential oil was isolated by a process similar to that for the preparation of *absolutes*, the yield being 0·016 per cent. It had a disagreeable and pungent odour, and only resembled the true odour of the flower after great dilution. The following materials were identified as constituents:

> Benzyl benzoate
> Benzyl alcohol
> Esters of cinnamic alcohol
> Hydrogen sulphide
> A body with a distinct odour of vanillin.

L. Hoejenbos and A. Coppens extracted the flowers of *H. orientalis, L.*, grown in the region of Amersfoort, Holland. They used benzene as the solvent and obtained a light brown product containing paraffins and waxes. These were separated by freezing an alcoholic solution and the odour-bearing constituents were isolated by steam distillation of the residue. The following constituents were identified:

> Benzyl alcohol   Dimethyl hydroquinone
> Phenylethyl alcohol   Eugenol
> Cinnamic alcohol   Methyl eugenol
> Cinnamic aldehyde   Benzyl acetate
> Benzaldehyde   Benzyl benzoate
> Methyl *o*-methoxy benzoate   Methyl-methyl anthranilate
> Ethyl *o*-methoxy benzoate   Benzoic acid
> Cinnamyl acetate

N-Heptanol and oenanthol were also probably present. Phenyl acetaldehyde and homologues, aliphatic aldehydes, alcohols and esters were not found.

Compounding notes

The reproduction of the hyacinth fragrance is one of the easiest in synthetic perfumery, as there are several definite chemical substances whose odour, particularly on dilution, has a marked resemblance to that of the flower. The most important of these is phenylacetic aldehyde, a liquid of syrupy consistency which has a tendency to polymerise and become exceedingly viscous, occasionally throwing down an unsightly deposit. It is known as *Hyacinthin*, and 4 to 5 grams dissolved in a litre of alcohol will produce the equivalent of a first pomade washing. Its odour is

Odour classification

Top notes	Middle notes	Basic notes
1. Amyl acetate	15. Cinnamyl acetate	65. Cinnamic alcohol
Benzaldehyde	Cinnamyl formate	79. Civet absolute
Benzyl acetate	Heliotropin	80. Hydroxy citronellal
2. Benzyl formate	Styrax oil	85. Phenyl acetaldehyde
Linalol	16. Eugenol	Dimethyl acetal
Phenyl ethyl acetate	18. Violet leaf absolute	89. Cyclamen aldehyde
Rosewood	21. Angelica seed	90. Galbanum resin
3. Dimethyl benzyl carbinol	Elemi resin	Opoponax resin
Terpineol	Ionone alpha	100. Acet eugenol
4. Citronellol	Methyl anthranilate	Amyl cinnamic aldehyde
Phenyl ethyl alcohol	22. Clove	iso-eugenol
6. Bergamot	24. Ethyl cinnamate	Musk ketone
Iso-butyl phenyl-acetate	Geranyl acetate	Olibanum resin
Phenyl methyl carbinyl acetate	Jonquille absolute	Phenyl acetic aldehyde
7. Benzyl propionate	Rose otto	Styrax resin
Methyl octine carbonate	Ylang	Undecylic aldehyde
8. Benzyl salicylate	30. Ambrette seed	Vetivert
Geraniol palmarosa	Cardamon	
Iso-butyl salicylate	31. Orange flower absolute	
Nerol	41. Mace	
Rhodinol	42. Acetyl iso-eugenol	
9. Tagette	43. Jasmin absolute	
10. Linalyl acetate	Rose absolute	
11. Galbanum oil	50. Laurinic alcohol	
Hyacinth absolute	Phenyl propyl aldehyde	
Narcissus absolute	Neroly	
Opoponax oil	60. Benzyl phenylacetate	
Sweet orange		
12. Methyl cinnamate		
Methyl heptine carbonate		
Petitgrain, French		
13. Paracresyl phenylacetate		
Elemi oil		
Phenyl propyl alcohol		
14. Mastic oil		
Methyl ionone		
Phenyl propyl acetate		

approximated to by *a*-bromstyrole, but this material has a rather harsh odour and is not used in fine perfumery. It finds employment, however, in the soap industry. Other aromatics having an odour reminiscent of hyacinth are cinnamic alcohol, styrax essence and galbanum oil, but whereas the alcohol may represent as much as 50 per cent of the finished compound, the essential oils should never exceed 5 per cent; and moreover, these figures are based upon the use of up to 10 per cent of phenyl acetic aldehyde. However, these materials alone can never yield an acceptable product; for while the absolutes of jasmin, rose, hyacinth, jonquille and narcissus are admirable flowering agents, it is necessary to look elsewhere for a solution to this problem. Opinions may vary, even among the experts, but it is safe to say that the key will be found to the simplest compounds in a combination of benzyl acetate, linalol, phenyl ethyl alcohol, heliotropin, ionone and one of the clove derivatives, where the crystalline synthetic may be employed in any quantity up to 20 per cent of the completed creation. While the above constituents are vital, innumerable permutations and combinations may be obtained by the skilful application of the odour classification, but even so it is surprising how pleasing variations can be secured by further additions of rose otto and hydroxy citronellal to adjusted percentages of the substances already noted.

Two formulae illustrate these suggestions:

Hyacinth, no. 1029

175	Benzyl acetate
100	Linalol
40	Phenyl ethyl alcohol
180	Heliotropin
10	Ionone
390	Cinnamic alcohol
5	Iso-eugenol
100	Phenyl acetic aldehyde
1000	

Hyacinth, no. 1030

150	Benzyl acetate
200	Linalol
100	Phenyl ethyl alcohol
10	Galbanum oil
20	Narcissus absolute

50	Heliotropin
40	Styrax essence
50	Eugenol
20	Ionone
10	Rose otto
30	Jasmin absolute
100	Cinnamic alcohol
200	Hydroxy citronellal
20	Phenyl acetic aldehyde
1000	

A finished perfume may be prepared thus:

Hyacinth, no. 1031

35	Compound, no. 1029
100	Compound, no. 1030
10	Ylang
5	Jonquille absolute
30	Ambergris tincture, 3 per cent
20	Musk tincture, 3 per cent
800	Alcohol
1000	

Jasmin

History

History does not appear to contain any definite references to the jasmin flower until about the sixteenth century. It belongs to the N.O. Oleaceae, and is supposed to be a native of India and to have become indigenous to southern Europe at an early date. Dioscorides tells us that the Persians obtained an oil from a white flower with which they perfumed their apartments during the repasts, and it is possible that this may have been the jasmin. The Hindus have always used perfumed flowers in the performance of their religious rites, and they also are stated to have had a particular preference for this flower. The origin of the introduction of jasmin into Italy has given rise to a legend recorded by Donald McDonald, which relates that a certain Duke of Tuscany was the first possessor of the plant, and as he wished to retain it as a novelty, forbade his gardener to give away a single sprig of it. The latter gentleman, however, seems to have neglected his instructions, and being an ardent lover, presented his lady with a bunch of the prohibited blossoms on her birthday. She was so charmed with the perfume of the flowers that she planted them in the ground, and by careful cultivation was able to produce large quantities of the blossoms, which she sold, and, after amassing a fortune, married the happy gardener. Today jasmin is so common that it is difficult to find many gardens in Europe that are not graced by its presence.

Varieties

The species most commonly known in Britain is *Jasminum officinale*, and its varieties have golden and silver-edged leaves, while some are double-flowered. It is a native of the warmer parts of Asia and became naturalised in the south of Europe at an early date. The blossoms appear from June to October, and while their perfume is sweet, its power does not approach that of the same species grown in warmer climates. *J. grandiflorum* is often grafted on to cuttings of this species.

J. sambac is a native of Arabia, but is found growing wild in India. It is a climber, with single and double white flowers which emit a most delightful fragrance. *J. trifoliatum* is a variety of this species known in India as *Kuddamulla* and in this country as the 'Tuscan' jasmin. It seems possible that this may be the variety referred to in the legend related above. The Indian ladies are said to hold the flowers of this jasmin species in great esteem and make them into necklaces. The blossoms are used for perfuming tea in China.

J. odoratissimum is a native of Madeira, with yellow flowers, which retain their perfume when dry. This species, known as 'Shuei flowers', is cultivated in Formosa where the blossoms are used for perfuming tea. They have been treated with volatile solvents by Tsuchihashi and Tasaki, who obtained 0·277 per cent of concrete oil, which upon maceration with alcohol yielded 0·116 per cent of volatile oil and 0·166 of floral wax. The oil was a reddish-brown liquid and contained the following constituents:

	per cent
Linalol	6
Linalyl acetate	6
Benzyl alcohol	1·6
Benzyl acetate	6
Indole	} 10
Methyl anthranilate	
Diterpene or sesquiterpene alcohol	57
Jasmone	nil

The flowers gave no result with enfleurage.

J. azonicum has white flowers, and is also a native of Madeira. It is sometimes grown in England.

J. paniculatum is a native of China, and has white flowers, which are used for scenting tea.

J. hirsutum has white flowers of exceptional fragrance, and is also a native of China and possibly India.

J. nudiflorum and *J. revolutum* are Chinese varieties with yellow perfumed flowers. The latter is found on the slopes of the higher mountains of Nepal, and is cultivated in India for the manufacture of perfume, as is also *J. auriculatum*.

J. gracillimum has white flowers.

J. grandiflorum is a native of the East Indies, and occurs in both double and single form. It is known as the Spanish or Catalonian jasmin, and its flowers retain their perfume when dried. The plant resembles *J. officinale*, upon which it is generally grafted in the south of France. *J. grandiflorum* is also grown in Tunis. The cuttings are placed about 2 feet apart and the blossoms collected for the preparation of perfume. In a well-kept plantation 1000 plants yield about 40 kilos of flowers.

J. gardeniodorum is indigenous to Togoland, where it flowers almost all the year round. It is a twining shrub having highly odorous white flowers, the perfume reminding of gardenias rather than of jasmin.

J. primulinum and *J. polyanthum* are two species recently found growing in Yunnan. They have large white (rose tinted on the edges) blossoms and are highly odoriferous.

J. Stephanense is a hybrid recently obtained by Javit by crossing *J. officinale var. grandiflorum* with *J. Besianum*. The plant is interesting on account of its having rose-coloured petals.

Odour

The odour of jasmin, like that of rose, is unique, and represents a type that cannot be *exactly* imitated at present by a mixture of any known synthetic aromatic chemicals or natural isolates. The discovery of the chemical structure of the jasmin ketone by two well-known makers of synthetics has brought the approximation of the natural odour by synthesis a step nearer. If the present high price of flowers continues, perfumers will be compelled to resort more and more to synthetic substitutes and in consequence a falling off in the demand for the natural product has already taken place. Many perfumes owe their fragrance to the skilful use of the jasmin odour, and it undoubtedly rivals that of rose for the premier position in a long list of aromatics.

Natural perfume

In the south of France the natural perfume is obtained from *Jasminum grandiflorum* which has been grafted on *J. officinale*. The flowers are white and highly odoriferous. Another species, *J. fruticans*, which has yellow flowers, has been experimentally grafted on this latter species. The resulting blossoms were white and extremely aromatic but the plantations thus grown died off very rapidly. The flowers are grown at Grasse in the following communes, and a few years ago yielded the weight of blossoms given below. But owing to the present high cost of collection little more than half these weights now reach the factories.

	kilos
Grasse—Montauraux	450,000
Mouans—Mougins	200,000
Pegomas—Cannes	100,000

The selection of ground for the cultivation of jasmin is very important. Water is necessary for its irrigation during the summer months and it must have a sunny aspect to avoid the frosts in winter. On the other hand, if the soil is too damp the plant dies, while if it is too warm the plant grows too quickly in the springtime. This likewise injures its growth. The ideal soil for plantation is that which has been sown with cereals for many years or has been grown to forage grops. Ground that has been planted with trees is used when necessary, but this is not without danger. If any roots are left in the ground and they decay, a disease develops, probably of a fungoid nature, which in Provence is called 'la mouffe'. This rapidly exterminates the jasmin. The most dangerous trees are fig, mulberry, and olive.

The ground is dug over deeply and then one of two methods of planting employed: either to plant directly or to transplant rooted plants. The latter is better, because more regular plantations result. The grafting is done a year later and some flowers are obtained after the first year. Some growers claim to have successfully grafted after six months. After blooming the jasmin is pruned and the ground covered to prevent freezing. In the following spring, when all fear of frost is over, the ground is uncovered and the plant shoots up. The blossoms appear generally about the middle of July and are collected up to the middle of October; in an exceptionally good year, collection may go on until the middle of November. The flowers gathered during August and September yield the largest quantity of perfume. They are picked off the stem as soon as possible after they open, which usually occurs in the evening up to the middle of August, and in the early morning after that date. Shortage of labour has made it difficult to adhere to these times. One hectare of jasmin, planted with 100 thousand cuttings, yields after grafting several hundred kilos of blossoms in the first year, and this rises to 4000 kilos in the fourth year. This yield is maintained until about the tenth year, after which it decreases. The cultivators of jasmin sometimes experience considerable losses through the attacks of caterpillars, which not only eat the leaves but attack the corollas about midway, with the result that the flower rapidly withers. The flowers after collection are immediately conveyed to the works, where their aromatic constituents are extracted by enfleurage or by means of volatile solvents.

The comparative value of these two processes in relation to this flower has led to much discussion and has already been referred to in the chapter on the production of flower perfumes. While Niviere does not consider there is any marked difference in the yield of *Essential Oil*, having regard to all the facts, von Soden has more recently determined the quantity of volatile oil obtained from the extraction of several thousand kilos of flowers, and he places the relative yield as between enfleurage and volatile solvents as 5 to 2. This chemist considers the latter process preferable because it recovers precisely those very important and as yet unknown odorous compounds which, owing to their lower degree of volatility, remain behind in the enfleuraged waste flowers with the enfleurage process, and are with difficulty absorbed into the fat. Flowers which remain lying in the chassis for twenty-four or forty-eight hours undergo radical changes while fading; new compounds arise in them and affect the recovery of odorous substances originally present in them which were left behind. E. S. Guenther made a number of experiments on these comparative processes, and whereas 700 kilos of jasmin flowers yielded 1 kilo of absolute by volatile solvents, only 225 kilos were necessary to produce the same weight of absolute by enfleurage. By steam distillation of the absolutes, the relative yield of volatile oil was $2\frac{1}{2}$ or 3 to 1 in favour of enfleurage.

At Avola in Sicily considerable quantities of jasmin are grown on the flower farms of UOP Fragrances, whose product is much appreciated at its competitive price.

In India jasmin flowers are extracted by a peculiar method of 'enfleurage'. Sesame seeds are washed, the husks removed and then dried. These are placed in layers alternating with jasmin blossoms and left thus for a period of twelve to twenty-four hours. Fresh flowers replace the exhausted ones until the 'pomade' becomes saturated. The oil is then expressed from the seeds and is known as 'siré ki tilli'.

In the 1959 edition of this work I mentioned the experimental jasmin plantations that were being developed in northern Africa, where formerly the cultivation of the plant on this continent had been confined to Egypt. Considerable areas are now under successful cultivation in Algeria and Morocco, and the resulting concretes and absolutes are available today from the Grasse houses that have sponsored this development. These sources are of particular importance at the present time owing to the extremely high and unrealistic prices of the Grasse products which, incidentally, are due solely to the higher costs demanded each year by the growers and pickers for the flowers delivered to the factory. While there is still a difference between the excellence of the fragrances from the four sources, the more reasonable prices of north Africa interest consumers, and has resulted in a decreased demand in Grasse and a corresponding increase in northern Africa.

Chemistry

There is some difference of opinion as to the constitution of the jasmin perfume, but it is generally recognised that the analysis made by Hesse and Muller represents its probable composition. A liquid otto of jasmin was obtained from pomade by steam distillation, and this was found to contain:

	per cent
Benzyl acetate	65
Linalol	15·5
Linalyl acetate	7·5
Benzyl alcohol	6
Other bodies	5·5

The composition of the unidentified constituents was investigated later by Hesse, and among those substances found were the following:

	per cent
Indole	2·5
Jasmone	3
Methyl anthranilate	0·5

Other materials identified include p-cresol and geraniol.

Hesse only obtained with certainty indole in the oil obtained by enfleurage, and he concluded that it did not exist in the living flower, but only in a complex form not revealed in the extracted or distilled oils. Thus indole, according to Hesse, was formed during enfleurage. This view has been rejected by von Soden and P. Baccarini, the former having found it in the oil obtained by volatile solvents and the latter (micro-chemically) in some species of jasmin. The question has been further investigated by Cerighelli whose experiments resulted in the following conclusions:

1. Indole is a normal constituent of *Jasminum grandiflorum*. In the flower it may exist as a complex combination not detected in the floral bud. When the flower opens the indole is liberated and is dissipated in the atmosphere. During the night indole accumulates in the tissues of the flower to be dissipated as the light acts on the plant.

2. Indole continues to be liberated by the gathered flowers; it only accumulates in a confined atmosphere.

3. Treated after keeping in a confined atmosphere (as is the case in industry), the flowers give up indole to the process of extraction and distillation.

4. Treated by enfleurage the flowers yield (over a period of twenty-four hours) three or four times more indole than by either extraction or distillation.

A further constituent, farnesol, has been identified by F. Elze and the presence of up to about 3 per cent of eugenol has been discovered by Mm. Sabetay and Trabaud.

The constitution of jasmone was studied by Treff and Werner in the laboratories of Heine, and they considered it to be a 3-methyl-2-(*n*-penten-2' yl)—cyclopenten-2-one-1 having the formula:

$$\underset{\underset{CH_2}{|}\quad\underset{CO}{|}}{\overset{\overset{CH_3}{|}}{\underset{CH_2}{\overset{C}{\diagdown}}}\diagup\overset{\diagdown}{C}-CH_2-CH=CH-CH_2-CH_3}$$

Similar discoveries were made by L. Ruzicka and M. Pfeiffer in the laboratories of Firmenich as far back as 1927 but were deposited with the Swiss Chemical Society and not published until 1933. This famous firm has never relaxed its intensive research into the composition of natural Jasmin absolute, and one of its most recent discoveries is methyl dihydrojasminate, which is now available under the name of Hediafone. It is a stable synthetic and free from dermatological risk; hence its use can be safely extended to cosmetic perfumes. From five to as much as twenty-five per cent can be employed in a variety of compounds.

All the above-mentioned bodies are prepared synthetically, and the newer Iso-Jasmone is now made by several Swiss and American houses; it has an intense odour reminiscent of celery and finds successful application in both Jasmin compounds and fancy perfumes.

Compounding Notes

The fragrance of the two inexpensive synthetics, benzyl acetate and amyl cinnamic aldehyde, so closely approximate that of jasmin that the duplication of its perfume is much facilitated by their use. And providing no natural absolutes are employed in the resulting creation its 'colourless' qualities make it particularly suitable for the perfuming of white cosmetic creams. The main problem is to correctly assess their respective proportions in relation to those of the other aromatics that are chosen to complete the blend; for the finished compound must be devoid of harshness and yet exhale the persistent sweet fruitiness that is typical of the flower. It is important to note that amyl cinnamic aldehyde should seldom be used in excess of 10 per cent of the whole, and the higher percentages require more careful treatment than the lower; so much so that it is often desirable to modify its odour with methyl ionone and one of the clove derivatives, or for preference a compound such as dianthus. The powerful 'lift' imparted by benzyl acetate requires to be correctly balanced as it is always tempting to use it to excess, but the figure of 30 per cent may be attained providing always the background is adequate. Phenyl ethyl alcohol and hydroxy citronellal are common constituents of these compounds, and linalol also may be added with advantage. Floral characters are imparted by ylang, and the finer the quality the better the results, while the overall fruitiness may

Odour classification

Top notes	Middle notes	Basic notes
1. Benzyl acetate	15. Cinnamyl acetate	65. Cinnamic alcohol
2. Benzyl formate	Heliotropin	79. Civet absolute
Linalol	21. Indole	80. Hydroxy citronellal
Phenyl ethyl acetate	Ionone alpha	85. Phenyl acetaldehyde
Octyl acetate	Methyl anthranilate	Dimethyl acetal
3. Paracresyl acetate	24. Ethyl cinnamate	87. Octyl aldehyde
Terpineol	Geranyl acetate	88. Ethyl methyl phenyl glycidate
4. Phenyl ethyl alcohol	Rose otto	89. Cyclamen aldehyde
7. Benzyl propionate	Ylang	91. Undecalactone
8. Amyl salicylate	31. Orange flower absolute	100. Amyl cinnamic aldehyde
Benzyl salicylate	43. Jasmin absolute	Benzoin resin
Linalyl propionate	Tuberose absolute	Decyl aldehyde
Nerol	50. Neroly	Iso-eugenol
Rhodinol	Phenyl propyl aldehyde	Liquidambar resin
9. Neryl acetate	59. Celery seed	Musk ambrette
10. Linalyl acetate	60. Benzyl phenylacetate	Musk ketone
11. Sweet orange		Phenyl acetic aldehyde
12. Petitgrain, French		Santal
13. Paracresyl phenylacetate		Styrax resin
Orris concrete		Tolu balsam
14. Methyl ionone		

be achieved by small additions of ethyl methyl phenyl glycidate or undecalactone. Finally there is a suspicion of narcissus in the flower, and one of the paracresol derivatives will provide it if used in small quantities. The inclusion of methyl anthranilate and indole are always problematical, and especially so the latter; for not only are colour changes inevitably associated with its use and ageing, but the compound may also develop a peculiar sourness which ultimately defeats its purpose.

The application of these principles is illustrated by the following formulae:

Jasmin Colourless, no. 1032

 200   Benzyl acetate
 100   Linalol

Jasmin Colourless, no. 1032 (*continued*)

200	Phenyl ethyl alcohol
5	Paracresyl phenylacetate
60	Geranyl acetate
10	Ethyl cinnamate
70	Ylang
300	Hydroxy citronellal
5	Undecalactone
50	Amyl cinnamic aldehyde
1000	

Jasmin, no. 1033

200	Benzyl acetate
100	Phenyl ethyl alcohol
5	Orris concrete
50	Methyl ionone
5	Indole
50	Ylang
30	Jasmin absolute
100	Benzyl phenylacetate
300	Hydroxy citronellal
100	Amyl cinnamic aldehyde
60	Dianthus compound
1000	

An example of a finished perfume follows:

Jasmin, no. 1034

90	Compound, no. 1032
45	Compound, no. 1033
6	Jasmin absolute
3	Orange blossom absolute
2	Rose absolute
1	Decyl aldehyde, 10 per cent
1	Civet absolute
2	Musk ketone
20	Ambergris tincture, 3 per cent
30	Musk tincture, 3 per cent
800	Alcohol
1000	

Lilac

The plant

There are three well-known species of lilac, *Syringa vulgaris*, the common lilac; *S. Chinensis*, the Chinese lilac; and *S. Persica*, the

Persian lilac. They belong to the N.O. Oleaceae, and are natives of the east. The common lilac is a bushy, erect shrub, native of Persia, and was introduced into this country by way of Turkey and Spain about 350 years ago. It has become one of our most common ornamental shrubs, of which there are several varieties, whose inflorescences may be red, purple, blue, or white. Of the *white* varieties the most popular are, Mari Legrange, Alba grandiflora, Alba magna, and Alba virginalis, while of the coloured varieties, souvenir de L. spath, stands supreme with its massive clusters of large blossoms.

The Chinese or Rouen lilac comes between *S. vulgaris* and *S. Persica*, and is also known as *S. dubia* and *S. rothomagensis*. The Persian lilac is distinct from the above two species as it is smaller and more erect. It was introduced into western Europe nearly 300 years ago. The branches are slender and spreading, and the flowers are deliciously fragrant. They take the form of small clusters, which are of a pale lilac colour, or nearly white. A pretty variety with deeply cut leaves, *S. luciniata*, is much esteemed.

During recent years several double forms have been introduced and are now much cultivated. The principal species are *S. Hyacinthiflora plena, S. Lemoinei,* and *S. Ranunculiflora*. The flower clusters are much denser and generally last longer than the single varieties.

There is a remarkable sequence of flowering in the lilac. The blossoms appear in France from 20-30 April; in Denmark from 1-15 May; in the south of Sweden and Russia from 1-15 June; in the centre of Russia, Sweden, and Norway after 16 June.

Winter lilac

Winter lilac, often seen in the florists during December and January, is one of the coloured forms which has been forced. The principal seat of this industry is around Paris at Fontenay-aux-Roses and Vitry-sur-Seine. Here are large sheds full of dry-looking stems which for about five years have been cultivated in the lilac fields outside. These stems are transported in trolleys along the rails that lead to the forcing chambers. The latter are situated on each side of a central alley, and look much like bathing cabins in public baths.

Each contains 200 lilac plants in a space of 36 by 12 feet. The stems are planted one by one in the soil which covers the ground to a depth of 6 inches, and the roof is glazed and fitted with adjustable blinds.

These forcing chambers are kept at a temperature of 86°F and curved pipes run along the walls arranged to impart the necessary dampness to the atmosphere.

Two days after their planting, the lilacs have already begun to bud, buds which are immediately suppressed, and only two to four flower buds are allowed later on to grow at the top of the stem, with probably two leaf-buds lower down for ornament.

The greatest care is taken during the opening of the flowers. Buds and leaves are mercilessly sacrificed, the thermometer is constantly consulted to see that the plant has the right temperature, from time to time the lilacs are sprayed with a vaporiser, and—most important item—as the flower approaches perfection the amount of daylight is increased, no light at all having been permitted till the petals began to open.

At the end of seventeen or eighteen days the blooms are ready to pick and this cutting generally takes place in the evening. The branches are taken to a cool cellar, and placed in water to preserve them. Finally they are tied in bunches round straw mushrooms and frequently surrounded by wallflower leaves. They are then carefully packed and despatched to the principal capitals of Europe.

Odour

The perfume of lilac in the form of the natural absolute is now an article of commerce, and is made by the progressive firm of Robertet after years of research by Maurice Maubert into the extracting of the flowers with butane. Pierre Dhumez was the first to attempt to solve the extraction problem with petroleum ether, but the odour of the resulting greenish-yellow oil was not even reminiscent of the blossoms. It was quite viscous in character but nothing is known of its constituents. G. Igolen extracted the fresh blossoms of *Syringa vulgaris* in the laboratories of UOP Fragrances. He used petroleum ether which gave 0·24 to 0·36 per cent of a dark green concrete. Benzene yielded 0·6 per cent of a blackish-green brittle concrete rich in waxes. The odour of both concretes was disagreeable. Thirty-eight per cent of absolute was obtained from the former with no improvement in odour. Steam distillation of this product yielded 8·72 per cent of oil, having a greenish-yellow colour and smelling of linseed oil. The presence of indole could not be confirmed.

The smell of lilac differs mainly between the coloured and the white types. The former have a fresher green character whereas the

Odour classification

Top notes	Middle notes	Basic notes

1. Acetophenone
   Benzaldehyde
   Benzyl acetate
   Ethyl acetoacetate
2. Linalol
   Octyl acetate
   Rosewood
3. Coriander
   Dimethyl benzyl carbinol
   Methyl acetophenone
   Phenyl Ethyl acetate
   Paracresyl methyl ether
   Paracresyl acetate
   Terpineol
4. Citronellol
   Phenyl ethyl alcohol
6. Bergamot
7. Geraniol, Java
   Benzyl propionate
8. Amyl salicylate
   Benzyl salicylate
   Geraniol palmarosa
   Rhodinol
10. Ethyl anisate
    Linalyl acetate
11. Decyl alcohol
    Linalyl cinnamate
    Sweet orange
12. Methyl cinnamate
    Petitgrain, French
13. Phenyl propyl alcohol
14. Basilic
    Phenyl propyl acetate
	15. Acet anisol
	    Cinnamyl acetate
	    Heliotropin
	    Styrax oil
	21. Anisic aldehyde
	    Indole
	    Ionone alpha
	22. Benzyl iso-eugenol
	    Phenyl cresyl oxide
	23. Methoxy acetophenone
	24. Anisyl acetate
	    Jonquille absolute
	    Rose otto
	    Ylang
	30. Ambrette seed
	32. Paramethyl hydro-cinnamic aldehyde
	43. Jasmin absolute
	    Rose absolute
	    Tuberose absolute
	47. Anisic alcohol
	50. Laurinic alcohol
	    Laurinic aldehyde
	    Neroly
	    Phenyl propyl aldehyde
	55. Nerolidol
		65. Cinnamic alcohol
		77. Methyl naphthyl ketone
		79. Civet absolute
		80. Dydroxy citronellal
		85. Phenyl acetaldehyde
		    Dimethyl acetal
		89. Cyclamen aldehyde
		91. Undecalactone
		100. Amyl cinnamic aldehyde
		     Benzoin resin
		     Coumarin
		     Iso-eugenol
		     Methyl nonyl acetaldehyde
		     Musk ketone
		     Olibanum resin
		     Peru balsam
		     Phenyl acetic aldehyde
		     Styrax resin
		     Tolu balsam
		     Vanillin
		     Vetivert

latter are extremely indoloid. They in fact remind very much of indole with the superimposition of hawthorn and jasmin. No white lilac perfume is perfect without indole.

Compounding notes

The finest grades of terpineol have an odour recalling that of lilac, but only so on extreme dilution. When the coloured flowers are smelled keenly it will be observed that the characteristic note has a perceivable fresh greenish background of rose, heliotrope and lily, and it is therefore convenient to consider these three flowers together in order to discover a suitable combination of aromatics which, when crowned with terpineol, will yield a duplication of the mother fragrance. Thus, geraniol and phenyl ethyl alcohol might well represent the rose; heliotropin and anisic aldehyde the heliotrope; and hydroxy citronellal and cinnamic alcohol the lily. When these materials are blended in suitable proportions it is only necessary to flower them with jasmin absolute in the more costly compounds, or with benzyl acetate in the cheaper products; enrich their bouquet with iso-eugenol; and impart a touch of green with phenyl acetic aldehyde. These, therefore, are the only vital constituents of a good pink lilac, but when the white variety is examined the presence of indole will be apparent, together with a fuller overall bouquet of jasmin and orange blossom. Thus, by varying the percentages of the already named aromatics for pink lilac, and by adding liberal quantities of neroly, plus the absolutes of jasmin and orange flower, an excellent white lilac may be created.

Two examples illustrate these suggestions:

Pink Lilac, no. 1035

35	Benzyl acetate
100	Terpineol
240	Phenyl ethyl alcohol
180	Heliotropin
10	Anisic aldehyde
130	Cinnamic alcohol
290	Hydroxy citronellal
5	Iso-eugenol
10	Phenyl acetic aldehyde, 10 per cent
1000	

White Lilac, no. 1036

1	Benzaldehyde
50	Benzyl acetate
25	Linalol
100	Terpineol
300	Phenyl ethyl alcohol
80	Petitgrain, French
40	Anisic aldehyde
5	Indole
30	Ylang
30	Jasmin absolute
20	Neroly
80	Cinnamic alcohol
200	Hydroxy citronellal
10	Phenyl acetaldehyde dimethyl acetal
9	Iso-eugenol
20	Phenyl acetic aldehyde
1000	

A lilac perfume may be prepared thus:

Lilac, no. 1037

110	Pink lilac compound
20	Muguet compound
1	Ambrette, 10 per cent
1	Cyclamen aldehyde
1	Methyl nonyl acetic aldehyde, 10 per cent
2	Rose absolute
5	Orange flower absolute
10	Jasmin absolute
40	Ambergris tincture, 3 per cent
10	Musk tincture, 3 per cent
800	Alcohol
1000	

Lily

History

Lilies, referred to in a general sense as beautiful white flowers, have been mentioned in literature as far back as 1014 B.C. In the Scriptures they are spoken of no less than eleven times, sometimes as a direct reference to the flower, as in The Song of Solomon ii 1 and 2, and occasionally as lily-work, for forming patterns of carved ornaments for the pillars and other parts of Solomon's temple, as in 1 Kings vii 19. Commentators have not been able to definitely clear up the botanical source of these flowers. John Smith says that *Lilium chalcedonicum* is the only true lily native of the Middle East, although the white lily, *L. candidum*, is much cultivated there, but is a doubtful native. Some suppose the first to have been the 'Lily of the Valley'[6] while Sprengel considers it to be the jonquille, *Narcissus Jonquilla*: others think it was *Amaryllis (Sternbergia) lutea*, an autumn-flowering bulb, with bright yellow flowers, a native of Europe and the Middle East, where it is abundant in the vales. It is, however, generally admitted that the lilies of the Bible cannot be identified with any special plant or plants, but that the term 'lily' is a general one for all plants having open lily-like flowers of showy colours, thus including *Anemone, Ranunculus, Cornflag* and even *Iris* which are abundant in the Middle East. Smith says that *Anemone coronaria*, with its brilliant colours, is the most conspicuous, and grows almost everywhere, without regard to soil or situation. It is abundant on the Mount of Olives and may well be considered to represent the 'lilies of the field' that surpassed 'Solomon in all his glory'. The plant familiarly known as 'Lily of the Valley' is not a native of the Middle East and cannot therefore be the plant of this name referred to in the Bible.

In ancient mythology the lily was held to be sacred and was consecrated to Juno, of whom it was said 'from the milk of her breasts sprang this beautiful flower'. Ancient writers have made many references to it. Theophrastus (400 B.C.) says: 'Krina show many variations in colour. ... The plant has in general a single stem, but occasionally divides into two, which may be due to differences in position and climate. On each stem grows sometimes one flower, but sometimes more; for it is the top of the stem which produces the flower, but this sort is less common. There is an ample root, which is fleshy and round. If the fruit is taken off, it germinates and produces a fresh plant, but of smaller size; the plant also produces a sort of tear-like exudation which men sometimes plant.' The lily was a favourite flower with Pliny who placed it second only to the rose. He says that it was much used for making an unguent called 'lirinon', and from his description of the whole plant there is very little doubt that it was *Lilium candidum* mentioned below. Pliny also describes a method of artificially producing crimson lilies from the bulbils of a white lily,

[6] Canticles ii. 1.

a proceeding altogether improbable and ridiculous. During this period the Greeks and Romans used the flowers at many of their feasts, and the former placed crowns of them upon the heads of their brides, as emblems of purity and abundance. There appears to be no record of the introduction of this bulbous plant into Britain, but it is assumed that the early Crusaders brought it from Palestine.

Varieties

The name *lily* is often given to pretty flowering plants in general, but in botany is restricted to the genus *Lilium*, of which there are many species, natives of the temperate zone of the Northern Hemisphere. They are showy flowering plants, favourites in gardens, and are represented by:

L. auratum, one of the most beautiful, some forms having flowers nearly 12 inches in diameter. The petals are broad and white with reddish-brown spots, and the centre is golden-yellow in colour. This species is native of Japan, and bulbs are imported annually.

L. bulbiferum is a handsome specimen about 2 feet high, bearing large crimson flowers shaded to orange.

L. candidum is well known as the white or madonna lily, and is universally cultivated. It is one of the finest examples of this genus, and its perfume when flowering in the summer is fragrant, strong, and honey-like.

L. croceum, also known as the crocus or orange lily, is a hardy plant bearing in early summer huge heads of rich, large orange flowers.

L. gigantium is a noble species native of Nepal. The stems are erect and from 6 to 10 feet high. They terminate with a huge raceme, measuring 1 to 2 feet, and having many long nodding fragrant flowers which are white, tinged with purple in the inside.

L. longiflorum is one of the most valuable garden varieties, and is known as the white trumpet lily, on account of its large trumpet-like blooms of snow-white purity and delightful fragrance.

L. tigrinum, the tiger lily, is probably the commonest kind in our gardens, and some varieties commence to flower at the end of August, while others are as late as October.

The two examples following are included here for the sake of completeness, as the synthesis of their perfume resembles that of the above very closely.

Lily of the Valley is probably the favourite member of the N.O. Liliaceae, on account of its exquisite fragrance. It is the single species of the genus *Convallaria*, known botanically as *C. majalis*, and on the Continent as *mugue* or *lys*. It is a native of Europe, found generally in shady places in woods, and frequently grown in gardens for its pure, waxy, odorous flowers which blossom from April to June. There is a variety with golden-striped foliage, and another with double flowers, but these are not so pretty. The finest form is called *Fontin's*, which is more robust than the common kind and develops larger flowers. This species is not the lily of the valley spoken of by King Solomon. The lilies of the valley which are sold by the florists about Christmas time were at one time 'forced' in Germany, but according to Gattefossé the industry has been developed in France since 1890 and is now a thriving one. The flowers are grown from wild rhizomes bearing latent flower buds called *turions*, and it takes about three years' cultivation before they are ready for forcing, although by special treatment successful results are sometimes obtained at the end of one year. Forced roots soon become exhausted and will not produce new shoots.

Belladonna Lily is one of the most important members of the N.O. Amaryllidaceae, and like the Guernsey lily, *A. sarniensis*, is a native of South Africa naturalised in the Channel Islands, from whence the bulbs are yearly imported to this country. The plant blooms late in the summer, and its flowers, which are of a delicate silvery rose colour, exhale a most delicious fragrance, recalling apricots. Other varieties, such as *A. treatea* and *A. cinnamonea*, are said to be natives of Brazil, and are largely cultivated in this country.

Odour

Different lilies, as we have mentioned, exhale slightly different odours, which in general may be described as delicate, elusive, and 'as sweet as honey'. The odour of ylang-ylang in *extreme dilution* is fairly representative of the type.

Natural perfume

This is extracted from *Convallaria majalis* by means of volatile solvents, and may be obtained as concrete or liquid absolute. The finest is that made by the firm of Robertet, who have improved the fragrance of the final product by extracting the flowers with butane. Only small quantities are produced, and in preparing

Odour classification

Top notes	Middle notes	Basic notes
1. Benzaldehyde Benzyl acetate 2. Linalol Phenyl ethyl salicylate Rosewood 3. Benzyl cinnamate Dimethyl benzyl carbinol Terpineol 4. Citronellol Phenyl ethyl alcohol 5. Ethyl phenylacetate 6. Bergamot Citronellyl formate 7. Benzyl propionate Geraniol, Java 8. Nerol Rhodinol 10. Linalyl acetate 11. Sweet orange 12. Methyl cinnamate Methyl heptine carbonate Petitgrain, French Phenyl ethyl cinnamate 13. Paracresyl phenylacetate Phenyl propyl alcohol 14. Methyl ionone	15. Cinnamyl acetate Citronellyl formate Heliotropin 18. Orris absolute 21. Anisic aldehyde Ionone alpha 24. Geranyl acetate Ylang 30. Cardamon 43. Jasmin absolute Rose absolute Tuberose absolute	65. Cinnamic alcohol 77. Methyl naphthyl ketone 79. Civet absolute 80. Hydroxy citronellal 85. Phenyl acetaldehyde Dimethyl acetal 87. Octyl aldehyde 89. Cyclamen aldehyde 91. Undecalactone 100. Benzoin resin Decyl aldehyde Musk ketone Santal Styrax resin Undecylic aldehyde

artificial lily perfumes the flowery note is generally obtained with other absolutes. Nothing is known concerning the composition of the lily odour. It is interesting to note, however, that Haensel extracted the leaves of *Convallaria majalis* by steam distillation and obtained 0·058 per cent of a semi-solid, greenish-brown volatile oil, having a pleasant aromatic odour.

Compounding notes

The most exquisite lily fragrance is undoubtedly that of muguet, or lily of the valley, and its perfume is the only one of this particular gamut that has ever achieved popularity. Since the odours of the other species of the genus lilium revolve round this note the following discussion will be devoted to it exclusively, and perfumers may thereafter adjust the other varieties to their own tastes.

Hydroxy citronellal is the one outstanding synthetic that can be said to truly duplicate the fragrance of the flower, and in consequence it occupies a prominent position in all muguet compounds, where it may occur up to as much as 50 per cent of the whole. Before its discovery the majority of lily perfumes were based upon terpineol and linalol, and these well known alcohols are still widely employed to modify the odour of the now accepted base. Another indispensable basic aromatic is cyclamen aldehyde, but its powerful and persistent fragrance limits its use in this delicate type to perhaps 5 per cent. Nevertheless, on carefully smelling the flower a rich rosy background will be readily perceived, and the choice for its reproduction herein rests between dimethyl benzyl carbinol, cinnamic alcohol and rhodinol. The latter possesses the most intense rosy sweetness and is capable of modification and even improvement by small additions of citronellyl formate, so that the use of all four substances in suitable proportions can be safely condoned in compounds of this class. Ionone is the only additional synthetic that may be subject to speculation, as in some compounds it is prone to develop a peculiar sourness with age. Should its inclusion be desired the quantity may with advantage be kept down to 5 per cent or less. The employment of cardamon oil is also problematical, and since the advent of cyclamen aldehyde it has fallen out of use.

A formula based upon these ideas follows:

Muguet, no. 1038

25	Benzyl acetate
30	Linalol
50	Dimethyl benzyl carbinol
20	Bergamot
20	Citronellyl formate
150	Rhodinol
40	Heliotropin
10	Ylang
150	Cinnamic alcohol
500	Hydroxy citronellal
5	Cyclamen aldehyde
1000	

A finished perfume may be prepared from the above compound and liberally flowered with absolutes, as indicated:

Muguet, no. 1039

120	Muguet compound
20	Pink lilac compound
4	Jasmin absolute
2	Rose absolute
2	Santal
1	Civet absolute
1	Musk ketone
40	Ambergris tincture, 3 per cent
10	Musk tincture, 3 per cent
800	Alcohol
1000	

Magnolia

The genus *Magnolia* is named in honour of a professor of medicine and botany, Pierre Magnol, who died in 1715. It consists of about sixty species of conspicuous trees and shrubs, natives of China, Japan, and America, and probably also of the mountainous districts of northern India. *M. grandiflora* appears to have been the first species to have been introduced into Europe, and was brought to France by the botanist Plumier about A.D. 1700.

Varieties

The most important species is the great laurel-magnolia, *M. Grandiflora,* a native of the southern United States. It sometimes attains a height of 80 or 100 feet, and is evergreen, with firm laurel-like leaves. The flowers are yellowish-white in colour, and stand upright, in the form of a cup 6 to 8 inches in diameter. They are highly odoriferous, and in England are the largest flowers of any trees growing in the open air. Their perfume is said to be inferior to that of the native-grown blossom, which has a predominating note suggestive of ylang-ylang backed up with orange blossom. There are several varieties of this species.

M. conspicua is a native of China known as Yulan, and as its name indicates, it is probably the most conspicuous species. It is deciduous, and attains a height of 40 to 50 feet, much branched and has pure lily-like flowers, produced before the leaves expand in such profusion that at a distance it appears one compact sheet of white.

Magnolia glauca is a native of the eastern United States, where it is known as swamp sassafras, sweet laurel, and beaver tree. It is a pretty, sub-evergreen shrub, about 20 feet high, with leathery leaves, which are bluish-green above and silvery below. The flowers are globular in shape, delightfully fragrant, and at first have a rich cream colour which gradually changes to a pale apricot tint. The leaves are fragrant and were examined by Rabak who obtained a yield of 0·05 per cent of a pale yellow aromatic oil. A cursory examination suggested the presence of phenols.

M. Kobus is a native of Japan and is a comparatively small tree, only attaining a height of about 10 feet. The bark has a camphoraceous odour, and the flowers are fragrant (suggestive of verbena). An oil distilled from the twigs has been examined by Asahina and Nakamura, who found the constituents mentioned below. This essence is known as Kabushi oil.

Other important species are *M. macrophylla,* with open, bell-shaped flowers, white, with a purple blotch at the base of the inner petals; *M. acuminata,* the cucumber tree, with lemon-yellow flowers; *M. tripetala,* the umbrella tree; *M. auriculata* and *M. cordata.* The botany of the various species is dealt with in detail by Sawer (*Odoragraphia,* II, 474).

Odour

The perfume of the majority of species of the *Magnolia* is exotic, and the fragrance resembles that of a ylang-lily complex, with a shading of clove and a top note of lemon.

Natural perfume

Oils have been obtained from Japan, but the species of magnolia from which they were extracted does not appear to be very clear. A list is appended of materials identified, but unfortunately they cannot be taken as a guide to the compounding of the synthetic perfume:

>Cineole
>Citral
>Anethol
>Eugenol
>Methyl chavicol
>Phellandrene
>Linalol
>Terpineol
>Caprinic acid
>Oleic acid
>And possibly pinene

G. Igolen extracted the flowers of magnolia in the laboratories of UOP Fragrances. The fresh blossoms, immediately after collection, were treated with petroleum ether, yielding about 1·5 per cent of a greenish-yellow concrete of brittle texture and melting at 58° to 60°C. Distillation with steam gave 10 per cent of a semi-solid essential oil of similar colour.

Compounding notes

The rich and exotic perfume of magnolia recalls, as we have seen, the ylang-ylang complex, shaded with a clove-like note and with an overall fragrance of lemon. A compound of this type could therefore be made most easily by the modification of an already created muguet. But since most perfumers prefer to work with raw materials the application of nerol in place of rhodinol should not be overlooked, nor that of the pepper-like property of eugenol which blends admirably with the constituents of muguet in the creation of this particular complex. Moreover, since the lemon

Odour classification

Top notes	Middle notes	Basic notes
1. Benzyl acetate	15. Guaiac wood	65. Cinnamic alcohol
2. Linalol	Heliotropin	80. Hydroxy citronellal
3. Coriander	16. Eugenol	89. Cyclamen aldehyde
Terpineol	19. Verbena	91. Undecalactone
4. Citronellol	21. Anisic aldehyde	100. Amyl cinnamic aldehyde
6. Bergamot	22. Benzyl iso-eugenol	Coumarin
8. Lemon	Geranyl formate	Decyl aldehyde
Nerol	24. Benzylidene acetone	Iso-eugenol
Rhodinol	Ethyl cinnamate	Musk ketone
10. Linalyl acetate	Geranyl acetate	Peru balsam
11. Sweet orange	Ylang	Tolu balsam
12. Petitgrain, French	27. Dimethyl hydroquinone	Vanillin
14. Methyl ionone	31. Orange flower absolute	
	43. Jasmin absolute	
	Rose absolute	
	Tuberose absolute	
	50. Laurinic aldehyde	
	Neroly	
	Phenyl propyl aldehyde	
	60. Citral	

note is of paramount importance the use of a fine quality citral is essential.

The following example illustrates the application of these points:

Magnolia, no. 1040

10	Benzyl acetate
30	Terpineol
60	Bergamot
150	Nerol
40	Heliotropin
50	Eugenol
100	Ylang
20	Jasmin absolute
10	Neroly
20	Citral

100	Cinnamic alcohol
400	Hydroxy citronellal
5	Cyclamen aldehyde
5	Decyl aldehyde, 10 per cent
1000	

A finished perfume may be prepared thus:

Magnolia, no. 1041

140	Compound, no. 1040
5	Orange flower absolute
2	Jasmin absolute
2	Rose absolute
1	Civet absolute, 10 per cent
20	Ambergris tincture, 3 per cent
30	Musk tincture, 3 per cent
800	Alcohol
1000	

Mimosa

History

Numerous species of *Acacia* (N.O. Leguminosae), known commercially as mimosa, are indigenous to Australia and certain parts of Africa. There is no clear record of their introduction to the southern part of Europe, but they appear to have been first cultivated in the neighbourhood of Cannes–Mandelieu–Vallauris about 1820, having been introduced from Australia. It has also been stated that the seeds were brought by Captain Ardisson from San Domingo in Haiti and that the plants were first grown in the region of Tanneron in 1839. Today there are over thirty varieties cultivated on the Côte d'Azur, where the cut branches are sold principally as cut flowers. Most of these blooms are 'forced', this industry having been started by Jacques Tournaire. The method used consists in subjecting the unopened flowers to the action of a humid atmosphere at a temperature of about 25°C, special apparatus being used for the purpose. Forcing takes from two to eight days according to season. The trade is so extensive that a special train leaves Mentone daily in January and February, and conveys to Paris large quantities of this favourite flower.

The principal centres now supplying mimosa for decoration

purposes are: Antibes, Biot, Cannes, Le Cannet, Frejus, Golfe-Juan, Mandelieu, Saint-Raphael, Théoule, Le Trayas, and Vallauris. Many plantations of other varieties have been laid out in the districts of Tanneron, Pegomas, Auribeau, Mandelieu, Croix des Gardes, California, and the Esterel. The newer varieties most favoured are known locally as 'Bon Accueil' and 'Motteana'. Visitors to the Riviera in February will have been charmed with the delicious fragrance and picturesque appearance of the mimosa trees as the train wends its way along the red rocky coast between Saint-Raphael and Cannes.

Varieties

All the mimosas have compact globose heads of fragrant yellow blossoms. The trees vary in height from about 18 to 30 feet. From the perfumer's point of view the following two species are of greatest value and most highly odorous:

Acacia floribunda whose perfume approaches that of cassie.

A. dealbata having a slightly coarser odour and resembling ylang-ylang.

In addition to these two the following species yield cut flowers for decorative purposes:

Acacia albicans or *Cultiformis*, with its bright golden-yellow flowers, forming some magnificent bunches of berries.

A. Trinervis, with cylindrical clusters of yellow flower chenilles, very pretty and original.

A. petialaris, of rapid growth and beautiful golden-yellow flowers.

Odour

The perfume of mimosa is faint (in small bunches) but charming, and resembles cassie, especially in *A. floribunda*, while in *A. dealbata* it distinctly recalls the odour of ylang-ylang. Several branches in a closed room give an overpowering fragrance.

Natural perfume

The mimosas from which the perfume is extracted come from the following districts: Auribeau, Biot, Cannes, Mougins, Pegomas, and Tanneron.

The flowers are collected mainly for sale as cut flowers from the middle of January to the end of March and by the middle of April this trade has practically ceased. It is, however, after this period that the tree flowers most freely when the blossoms are more highly scented and open, and in consequence are received at the Grasse works in greater quantities where they are extracted by the volatile solvent process (distillation and maceration yield unsatisfactory and inferior products), the yield of concrete being from 0·7 to 0·8 per cent from which about 20 per cent of absolute is obtained (equivalent to about 9 per cent of colourless absolute). The product of *A. floribunda* is a viscous, oily liquid, while that of *A. dealbata* contains a higher percentage of wax and is concrete. Both absolutes have an intensely sweet honey-like odour, reminding of orris, are comparatively cheap and give excellent results in floral Colognes (*Eau de Cologne*), while in 'de luxe' perfumes, particularly honeysuckle and heliotrope, they give floral notes which are unique: very fine bouquets of honey-like odour can be obtained by combination with muguet or jasmin.

Chemistry

Very little is known of the chemistry of these oils, though H. von Soden distilled with steam an absolute prepared with volatile

Odour classification

Top notes	Middle notes	Basic notes
2. Linalol	15. Guaiac wood	65. Cinnamic alcohol
Methyl salicylate	Heliotropin	73. Cassie absolute
3. Methyl acetophenone	21. Anisic aldehyde	77. Methyl naphthyl ketone
Terpineol	Ionone alpha	79. Civet absolute
	Methyl anthranilate	
4. Phenyl ethyl alcohol	24. Benzylidene acetone	80. Hydroxy citronellal
6. Bergamot	Ylang	100. Amyl cinnamic aldehyde
8. Nerol	43. Jasmin absolute	Decyl aldehyde
9. Methyl anisate	Rose absolute	Iso-eugenol
10. Linalyl acetate	50. Laurinic alcohol	Musk ketone
12. Methyl heptine carbonate	Laurinic aldehyde	Peru balsam
13. Orris concrete	Phenyl propyl aldehyde	Phenyl acetic aldehyde
14. Mimosa absolute	Undecylic alcohol	Tolu balsam

solvents and, calculated as fresh flowers, obtained 0·018 per cent of yellowish-green oil, which solidified in ice into a mass of crystalline leaflets.

More recently mimosa perfume has been studied by Sabetay and Trabaud, who identified the following constituents: anisic acid, palmitic acid, acetic acid esters, palmitic aldehyde, a primary alcohol, and large quantities of unsaturated aliphatic hydrocarbons.

Compounding notes

The relatively low cost of natural mimosa might lead one to believe that it would be widely used, but for some strange reason its fragrance has never appealed greatly to the public, and in consequence synthetic flower oils of this type suffer from a similar lack of interest. Nevertheless, the duplication of the odour is worthy of study, and falls into the same category as those of acacia and cassie, but is finer than the former and not so fragrant as the latter. Thus, the problem revolves round either anisic aldehyde, the alcohol and/or its esters, one or other of which must be blended with other aromatics that will develop and modify it sufficiently well to produce a satisfactory creation.

Let us therefore examine the perfume of the flower; first by smelling a nosegay of the lovely blossoms, and afterwards the bouquet they exhale in a closed room. We shall immediately notice that the powerful anise background is masked by an overall lilac of greenish tonality, and with perhaps a suspicion of orange blossom; qualities of odour that are most apparent in the diffused perfume. Thus, after selecting the most suitable lilac constituents, of which terpineol is most important in this case, the main problem resolves itself into obtaining a good balance between them and possibly anisic aldehyde. Once this has been done, it is only a question of imparting a touch of green and of an orange type synthetic, and finally flowering the whole with natural mimosa, cassie and jasmin to the utmost limit allowed by costs. While methyl anthranilate or methyl naphthyl ketone will provide the suggestion of orange blossom, the greenish tonality is a matter of choice and lies between methyl acetophenone, methyl heptine carbonate and phenyl acetic aldehyde, or for preference the first two in smallish quantities.

A mimosa compound may therefore be built up on these lines:

Mimosa, no. 1042

4	Methyl acetophenone
450	Terpineol
50	Phenyl ethyl alcohol
20	Bergamot
1	Methyl heptine carbonate
20	Mimosa absolute
30	Heliotropin
180	Anisic aldehyde
5	Methyl anthranilate
30	Jasmin compound
100	Cinnamic alcohol
10	Cassie absolute
100	Hydroxy citronellal
1000	

A finished perfume may be prepared thus:

Mimosa, no. 1043

140	Compound, no. 1042
5	Mimosa absolute
3	Cassie absolute
1	Jasmin absolute
1	Musk xylene
20	Ambergris tincture, 3 per cent
30	Musk tincture, 3 per cent
800	Alcohol
1000	

Narcissus

History

There are botanical references to this beautiful genus in the works of Theophrastus, but it is to Ovid that we owe the story of its creation. In his *Metamorphoses*, Book III, he tells us that Cephisus, the river god, bore a beautiful child named Narcissus, and when Tiresias, the seer, was asked if he would live to a ripe old age he replied, 'If he ne'er know himself'. Mirrors were kept from him, and the saying of the prophet seemed but empty words, for Narcissus had reached the age of sixteen, and many maidens sought his love. But he was proud and cold, and so slighted the affections of Echo, the nymph, that she hid in the woods, became gaunt and faded until only her voice remained. Narcissus went to a pool one day to quench his thirst, and while there another thirst sprang up, for while he drank, he was smitten by the sight of the beautiful form he saw reflected in the water and fell passionately in love with it. He attempted to clasp it in his arms, but after fruitless efforts killed himself in desperation. When he had been received into the infernal abodes, he kept on gazing at his image in the Stygian pool. His naiad sisters beat their breasts and shore their locks in sign of grief for their dear brother. When they prepared the funeral pile, the torches, and the bier, his body could not be found, but in its place was a flower, its yellow centre girt with white petals. That flower into which he had been changed still bears his name! Pliny does not agree with this fable, but states[7] that the herbacious narcissus[8] produces dull, heavy pains in the head and in consequence derived its name from 'Narce' (torpor or lethargy). He mentions three varieties of flowers and in another book[9] speaks of an emetic bulb which commentators have identified as that of *N. Jonquilla*.

The narcissus is stated to have been brought to this country by the Romans.

Varieties

Narcissus is the botanical as well as the familiar English name for a genus of bulbous plants of the N.O. Amaryllidaceae, of which there are about twenty species. The following are worthy of note:

N. bulbocodium is a species having slender, rush-like leaves, and is known as the hooped petticoat daffodil. In Spain it flourishes in wet meadows during the winter and spring, but disappears at other times. Here the colour is golden-yellow, while in southern France it is sulphur-yellow.

N. Jonquilla is known as the rush daffodil or jonquille, and is often found in our gardens, the bulbs being imported from Holland and Italy for forcing in pots. This fragrant plant is one of the most powerfully scented of all the genus, and a few blooms are sufficient for perfuming a large room.

N. Poeticus, poet's or pheasant's eye narcissus, is one of the oldest and most charming varieties, being widely distributed in

---

[7] Book XXI, chap. 75.   [8] *N. pseudo-narcissus.*   [9] Book XX, chap. 41.

Europe from the Pyrenees to Germany. It flowers from April to June. *Gardenia narcissus* is an elegant double variety grown largely for the London markets.

N. Pseudo-narcissus, the common daffodil, of which there are hundreds of varieties, both growing wild and under cultivation. They are divided into three principal kinds, golden, bicolours, and sulphur and white.

N. tazetta, the polyanthus or bunch-flowered narcissus, is a native of southern Europe and western Asia, being abundant in Palestine, and during the flowering season it is to be found in nearly every house, especially in Damascus. By some it is considered to be the rose of Sharon, the original Hebrew word bulb being translated rose, and indeed a rosebud is something similar to the bulbs of this plant. It is the classic narcissus of Homer, Ovid, and other poets, both Greek and Roman. In China the flower is held to be sacred, and is known as the joss flower or sacred lily. The plant from each bulb bears tall, many-flowered, charming lily heads of from six to twenty-four large yellow or yellowish-white flowers. They are powerfully fragrant and recall the odour of jonquille.

The majority of the narcissi and jonquilles sold by the florist in London during March come from the Scilly Isles. Here a large proportion of their income is derived from flower growing, and especially so at St Mary's where the blossoms are cut before they open, the forcing being carried out in steam-heated pits.

Natural perfume

In the south of France the cultivation of both narcissus and jonquille had fallen off until a few years ago. The increased demands of the perfumer for the natural perfume of these two flowers gave some encouragement to the cultivator, who had found it scarcely worth while to pay any attention to them owing to the poor price paid for the blossoms. In 1970 the harvest yielded about 53,100 kg of narcissus and about 5000 kg of jonquille. The principal centres for the cultivation of the former are: Caille, Cogolin, Grasse, Magagnosc, Hyères and Tourettes; and for the latter: Grasse, Peymeinade, and Tanneron. Plantations of jonquille as well as important improvements in the old ones have been made in the communes of Peymeinade, Tanneron, Callian and Montauroux. A large quantity of the narcissus is sold as cut flowers for decorative purposes. It is known locally as '*Done*'.

The principal species cultivated for perfumery purposes is *N. Tazetta*. Nowadays *N. Poeticus* is mainly used, since it grows fairly profusely in the higher regions around Grasse and the only cost is that of transport. The crop is collected during April and May. *N. odoris*, Willd., and *N. Joncifolius* are also grown but are generally left aside since their perfume is weak and suggestive of jonquille. The narcissus is extracted by volatile solvents and the yield of concrete is about 0·3 per cent which in turn gives about 30 per cent of absolute.

The jonquille flowers yield a much more powerful perfume. The bulbs are planted out in rows in the autumn and towards the end of March the blossoms appear—four or five bright yellow flowers on each stem. Each bloom is cut off separately and a woman will collect from 400 grams to 1 kilo in an hour according to the density of the flowers. If good results are desired the bulbs have to be replaced every two or three years. Jonquille flowers are extracted by volatile solvents and yield about 0·3 per cent of concrete which in turn gives over 40 per cent of absolute. They are used in the proportion of 3 of flowers to 1 of fats when extracted by maceration. The perfume is also isolated by means of enfleurage. In Holland large quantities of *N. Poeticus* are now grown for perfumery purposes and the blossoms are extracted by the volatile solvent process. The indoloid character of the absolute makes its use in fine perfumes rather risky owing to discoloration. This is minimised by the employment of the decolorised product—a fine, flowery and powerful asset.

Chemistry

The flowers of yellow narcissus were extracted with volatile solvents by H. von Soden who then distilled the product with steam. Calculated as fresh flowers, he obtained 0·0068 per cent of an oil having an odour of narcissus, so narcotising as to cause headache. This chemist treated jonquille flowers in a similar manner and obtained a yield of 0·1577 per cent of a colourless oil which rapidly became yellowish-brown on exposure to air. A chemical examination indicated the presence of the following substances: methyl and benzyl benzoate, indole, methyl anthranilate, esters of cinnamic acid—particularly the methyl ester and probably also linalol. A further investigation of the jonquille perfume was made by Elze who suspected the presence also of

FLOWER PERFUMES: NARCISSUS

eugenol, geraniol, nerol, farnerol and jasmone. The chemistry of these two flower oils is thus still comparatively vague, but judging by the odour the following substances might also be present:

>Benzyl acetate
>Para-cresol or derivatives
>Phenylethyl esters

Odour classification

Top notes	Middle notes	Basic notes
1. Benzyl acetate	15. Acet anisol	65. Cinnamic alcohol
2. Linalol	Heliotropin	77. Methyl naphthyl ketone
Phenyl ethyl acetate	Rose otto	
	Styrax oil	79. Civet absolute
3. Paracresyl acetate	16. Eugenol	80. Hydroxy citronellal
Paracresyl iso-butyrate	21. Indole	
	Ionone alpha	87. Octyl aldehyde
Terpineol	Methyl anthranilate	90. Rhodinyl acetate
4. Citronellol	22. Orange flower water absolute	91. Phenylethyl phenylacetate
Methyl methyl salicylate	Phenyl cresyl oxide	Undecalactone
Phenyl ethyl alcohol	24. Geranyl acetate	100. Acet eugenol
	Jonquille absolute	Amyl cinnamic aldehyde
6. Bergamot	Ylang	
7. Benzyl iso-butyrate	30. Ambrette seed	Benzoin resin
	31. Orange flower absolute	Costus
8. Benzyl salicylate		Coumarin
Geraniol palmarosa	42. Acetyl iso-eugenol	Decyl aldehyde
Nerol	43. Jasmin absolute	Iso-eugenol
Rhodinol	Rose absolute	Labdanum resin
9. Methyl anisate	Tuberose absolute	Methyl nonyl acetaldehyde
10. Linalyl acetate	50. Laurinic aldehyde	
Phenyl propyl propionate	Phenyl propyl aldehyde	Musk ketone
		Musk xylene
11. Narcissus absolute	Neroli	Peru balsam
12. Methyl cinnamate	60. Benzyl phenylacetate	Phenyl acetic aldehyde
Petitgrain, French		
Phenyl ethyl cinnamate		Santal
		Styrax resin
13. Orris concrete		Vanillin
Paracresyl phenylacetate		
Phenyl propyl alcohol		

Compounding notes

The penetrating fragrance of Narcissus has been likened to a blend of jasmin with hyacinth, but experienced perfumers will have noticed the peculiar cresol/indole note that dominates the whole, together with a touch of clove in the background. It is therefore relatively easy to arrive at the floral complex in synthetic prototypes, whereas the real key to their true character lies in the choice and blending of the above indispensable constituents and the ratio between them and the floral aromatics. At one time paracresyl acetate was favoured for this purpose, but it is not satisfactory owing to the unpleasant smell it imparts to the whole on ageing. It is therefore safer to employ the phenylacetate as it is more stable and as much as 7 per cent may be included without fear of deterioration. The quantity of indole is a matter of taste, but since it is prone to discolour on standing the lower the percentage the better. Iso-eugenol admirably provides the deep clovy note, while the suggestion of hyacinth may be obtained with phenyl acetic aldehyde. The importance of nerol as a top note constituent should not be overlooked.

A good compound may be prepared as follows:

Narcissus, no. 1044

200	Benzyl acetate
200	Linalol
50	Phenyl ethyl alcohol
150	Nerol
70	Paracresyl phenylacetate
1	Indole
70	Ylang
100	Cinnamic alcohol
100	Hydroxy citronellal
10	Amyl cinnamic aldehyde
40	Iso-eugenol
9	Phenyl acetic aldehyde
1000	

A perfume may be prepared from the above thus:

Narcissus, no. 1045

140	Compound, no. 1044
4	Narcissus absolute
3	Jasmin absolute
2	Rose absolute

1	Civet absolute
30	Ambergris tincture, 3 per cent
20	Musk tincture, 3 per cent
800	Alcohol
1000	

New-mown hay

History

Hay is not a charming thing to look upon and does not therefore appear to have inspired the ancient poets. The word, however, occurs in the early scriptural writings, but it seems doubtful whether the reference is to cut hay as we know it today. Thus in Isaiah:[10] 'The hay is withered away, the grass faileth, there is no green thing.' Again in Proverbs:[11] 'The hay appeareth, and the tender grass showeth itself.'

In both cases the inference is that the plants are growing. Furthermore, in the Middle East the cutting, drying, and stacking of grass does not appear to be practised.

Pliny devotes a whole chapter[12] to details concerning haymaking, but there is no reference to the odour so noticeable while the plants are drying.

The plants

The peculiar fragrance of freshly cut hay is at once charming and unique. Many fodder grasses grow spontaneously in our pastures, but several are especially cultivated and include *Lolium perenne, L. italicum, Pheleum pratense, Poa pratense, Cynosurus cristatus* and *Anthoxanthum odoratum*. The latter, also known as sweet-scented vernal grass, contributes more towards the characteristic odour of the hayfields than the others, and peculiarly enough the odoriferous constituents are only developed in the stem when dried. Among other plants in which the hayfields perfume is noticeable are the following:

Alyssum compactum. Sweet alyssum bears small white flowers, and the plant is particularly fragrant after a shower of rain.

Asperula odorata, woodruff, is a wild perennial of the N.O. Rubiaceae, growing in shady places near woods. The plant flowers

---

[10] xv. 6; 726 B.C. [11] xxvii. 25; 700 B.C. [12] Book XVIII, chap. 67.

in May, and the small white blossoms are profusely distributed over the tufts of whorled leaves. The odour is emitted by the stem and leaves when dry, and in Germany these are put into wine to give it a distinctive flavour. There are other varieties with rose-blue flowers.

Liatris odoratissima is an evergreen plant thriving in the swamps of North America, where it is known as the vanilla plant, although it has no connection with the vanilla of commerce. The leaves are disagreeable when fresh, but develop a delightful fragrance, resembling tonka bean, when dry. They are used for flavouring cigars. Coumarin was at one time extracted from the leaves of this species.

Liatris spicata is also a native of North America and belongs to the N.O. Compositae. Some years ago it was cultivated at Marseilles by M. Davin and found useful for perfuming tobacco, linen, etc. Apparently it contains less coumarin than the above-mentioned species and is therefore not so valuable.

Melilotus officinalis, the common melilot, is a clover-like annual or biennial of the N.O. Leguminosae, and when withered by the sun is almost as fragrant as the woodruff.

Melilotus alba, known as Bokhara clover, is used chiefly for bee-feeding on account of the honey found in the white blossoms. The seed and dried leaves are perfumed.

The natural perfume is due to the presence of coumarin, which is formed while drying by the action of a ferment on a constituent of the plant.

Chemistry

Bourquelot has shown that this constituent is probably d-glucose and the ferment probably emulsin.

Some peculiar facts concerning the liberation of the coumarin were brought to light by Heckel. In the course of experiments with anaesthetics he was able to show that chloroform and ethyl ether immediately set free this lactone from the leaves of *Liatris spicata*. In the case of *Anthoxanthum odoratum* similar liberation only took place after ten minutes, but when a mixture of methyl and ethyl chlorides was employed, the separation of the coumarin was effected at once. Similar experiments were conducted with *Melilotus officinalis*.

Odour classification

Top notes	Middle notes	Basic notes
1. Acetophenone Benzyl acetate	15. Acet anisol Heliotropin	79. Civet absolute 80. Hydroxy citronellal
2. Linalol Methyl salicylate	20. Clary sage 21. Anisic aldehyde Ionone beta	100. Benzoin resin Benzophenone
3. Cuminic aldehyde	24. Ylang	Coumarin
4. Lavender	27. Dimethyl hydroquinone	Methyl nonyl acetaldehyde
6. Bergamot	29. Geranium bourbon	Musk xylene
7. Geraniol, Java	31. Orange flower absolute	Oakmoss
8. Amyl salicylate Rhodinol	43. Jasmin absolute Rose absolute	Patchouli Santal
10. Diphenyl oxide	50. Laurinic aldehyde Neroli	Tolu balsam Tonka resin Vanillin Vetivert

Compounding notes

Everyone loves the powerful fragrance of new-mown hay, and it is one of the few perfumes that will realistically recall to the male those places where it has been encountered. I have enjoyed it in many parts of the world, but in none so delightfully as the wild Val di Fassa, in the Dolomites. I had set out one sunny summer morning to walk up to the Contrin Hut, where I was to meet my guide whence we were to begin the arduous roped ascent of Marmolada, the highest peak in this group of spectacular mountains, on the following morning at 6 a.m. if weather permitted. Wandering along slowly, with a forty pound rucksack on my back and with an ice axe in my hand, I entered the last stretch of the valley before mounting the rough track through the pines; and there I passed through fields where the hay had been cut on the previous afternoon. Its perfume was marvellous and I could only account for its intensity and remarkable fragrance by the heat of the morning in such a steeply enclosed valley. I shall never forget it and every time I smell new-mown hay my mind travels like lightning to that unforgettable scene. The duplication of the odour is one of the easiest in synthetic perfumery, because it is dominated by that of coumarin which is therefore an indispensable raw material. The only other 'musts' are linalol and lavender; for thereafter the perfumer must exercise his ingenuity to complete the blend. The suggestions incorporated in the following formula may be modified to taste in accordance with the odour classification.

Foin coupe, no. 1046

70	Acetophenone
70	Benzyl acetate
300	Linalol
150	Lavender
40	Bergamot
20	Clary sage
50	Geranium bourbon
50	Benzophenone
200	Coumarin
20	Musk xylene
5	Oakmoss
10	Patchouli
15	Santal
1000	

A finished perfume may be prepared as follows:

New-mown hay, no. 1047

130	Compound, as above
3	Orange flower absolute
4	Jasmin absolute
2	Rose absolute
1	Civet absolute
5	Tonka resin
25	Musk tincture, 3 per cent
830	Alcohol
1000	

Orange blossom

History

The orange tree is very probably a native of northern India, but there is no definite indication of the date it was brought into the Western Hemispheres, although this is believed to be about the ninth century. There is one reference in the Scriptures[13] which

[13] Prov. xxv. 11.

reads—'a word fitly spoken is like apples of gold in pictures of silver'. From this, some have presumed that oranges were referred to, but according to different commentators the fruit was not cultivated in Palestine at the time of Solomon. The orange is not mentioned by the early Greek and Roman writers and it must be presumed therefore that it was unknown to them. The Arabs seem to have brought the trees from India, first to Africa, Arabia, and Syria and later to Italy, Spain, and Sicily. Avicenna, an Arabian doctor, who flourished at the commencement of the tenth century, appears to have employed the juice of the orange medicinally.

There is evidence to show that the orange first cultivated was the bitter orange. The orange tree at Rome, said to have been planted by St. Dominic, A.D. 1200, and which still exists at the monastery of St. Sabina, bears a bitter fruit, and the ancient trees standing in the garden of the Alcaza at Seville are also of this variety. These authorities also state that the sweet orange began to be cultivated about the middle of the fifteenth century, having been introduced from the East by the Portuguese. It has probably long existed in southern China, and may have been taken thence to India. One of the first importations of the fruit into Britain was in A.D. 1290 when the queen of Edward I bought some of the cargo of a Spanish ship at Portsmouth.

Orange blossoms have for centuries been worn at weddings. The origin of this custom is attributed to the Saracens, whose brides used to wear orange blossom as a sign of fecundity. The custom was introduced to Europe by the Crusaders.

The mythological conception of the fruit has always been associated with the hymeneal altar. The orange is by many supposed to be the golden apple presented by Jupiter to Juno on the day of her nuptials. The apples could be preserved nowhere but in the garden of Hesperides, where they were protected by three nymphs bearing that name, the daughters of Hesperus, and by a more effectual and appalling guard, a never-sleeping dragon. It was one of the labours of Hercules to obtain some of these golden apples. He succeeded, but as they could not be preserved elsewhere, it is said they were carried back again by Minerva.

As far back as the sixteenth century the Italian philosopher Porta obtained what he described as an oil of exquisite fragrance by distilling flowers of the citron tribe. About 1680 the oil from orange flowers was called essence of neroli from the fact that it was used for perfuming gloves by the wife of Flavio Orsini, prince of Neroli.

The orange trees are now much cultivated in Europe for both their flowers and fruit. In China, and particularly at Canton, the blossoms are collected when fully expanded and are used for scenting tea known as Orange Pekoe.

Varieties

Citrus aurantium is a low, much-branched tree of the N.O. Rutaceae. It attains a great age, and in some parts of Spain is stated to be 600 years old, while one growing in a box at Versailles is said to have been planted in 1421. In Spain, Portugal, and Italy the plants are cultivated for their fruit, but in the south of France they are grown almost exclusively for their flowers. Here they are grown from pips, and the young plants are grafted when about three years old. When the seedling is about 4 feet high it is transplanted and allowed a year to gain strength before being grafted. The cultivation of the plant requires much attention, and a crop of flowers is not expected before the fourth year after transplantation.

In a good year an orange tree will yield from 12 to 15 kilos of flowers and these are collected principally by women and children.

The total crop was about 2 million kg, but this figure has been exceeded in the last decade.

At Boufarik in northern Africa the orange tree is also much cultivated.

The sweet or Portugal orange is the *Citrus aurantium* of Risso, who enumerates nineteen varieties.

The bitter or Seville orange is the *Citrus Bigaradia*, Duhmel; *C. aurantium*, var. *amara* Linnaeus; and *C. vulgaris* of Risso, who describes twelve varieties. These are, however, in every respect superior to the former kind.

Odour

The flowers of both sweet and bitter varieties are exceedingly fragrant, although those of the latter have a finer aroma, which recalls jasmin very slightly. The trees of the orange grove all blossom about the same time, and the perfume is perceptible at a considerable distance.

Natural perfume

The bitter orange tree yields four distinct products which are of value to the perfumer. They are:

Oil of neroli, from the blossoms by distillation.

Orange flower absolute, from the blossoms by extraction (warm fats or volatile solvents).

Oil of Petitgrain, from the leaves and twigs by distillation.

Oil of orange, from the peel by expression.

In the south of France the orange flower crop is collected during May (a second crop is obtained in October if conditions are favourable), and the yield of oil depends upon atmospheric conditions and upon the time of collection. The best results are obtained from the middle of the month onwards, when if the weather be warm and dry 1000 kg of flowers will produce as much as 1500 grams of oil, although higher figures have been recorded. The average is nearer 1000 grams, and under adverse conditions may be as low as half that amount.

In Tunis there are large orange groves from which the flowers are collected for the exclusive purpose of distillation. This oil is much appreciated and in price compares favourably with the Grasse product. Algerian neroli has not been available for some years owing to the political situation.

Charabot and Laloue studied the formation of neroli oil in the orange blossom and found that the quantity increased during the flowering period; dry flowers yield a higher percentage of oil than fresh. The formation and accumulation of oil is more active when the blossom is fully developed and the ester content increases. The geraniol content increases, while the linalol diminishes. These chemists observed no marked difference in the constitution of the oil derived from the petals alone and that of the other floral organs, but the petal oil always contained slightly more methyl anthranilate.

Jeancard and Satie observed that in the distillation of orange blossom the yield of oil during harvest-time increased in good weather towards the end of May, but diminished on rainy days.

As mentioned above, orange blossom is collected in October as well as May and a comparative examination of the two products was made by Roure-Bertrand Fils. In each case they extracted 300 kilos of flowers with petroleum ether, the spring flowers yielding

0·2272 and the autumn flowers 0·1795 per cent of concrete. The essential oil was isolated by steam distillation. The direct oil and that dissolved in the water were examined separately, with the following results (abridged):

	Oil collected in the receiver		Oil dissolved in the water		Total yield	
	May	Oct.	May	Oct.	May	Oct.
	per cent	per cent	per cent	per cent	per cent	per cent
Yield	0·0659	0·0611	0·0077	0·0052	0·0736	0·0663
Esters (as linalyl acetate)	25	33·8	21·5	20·6	24·6	33·4
Alcohols	54	59	51	54	52	57
Methyl anthranilate	2·4	1·5	12·2	13·4	3·5	2·7

The lower yield of oil in the autumn blossoms is accounted for by a generally lower temperature and by the more active circulation of the constituents in the plant during the spring. The difference in the composition of the two oils is ascribed by Roure-Bertrand Fils to the fact that in spring the young shoots determine the constituents while in autumn the maturer branches function in this manner. The low methyl anthranilate content is unexplained, since it is usually 7 per cent, but may be accounted for by the very severe winter which preceded these experiments.

Although, as stated above, the blossoms may be treated by different methods for the extraction of their odoriferous constituents, that of distillation is most common and represents approximately 80 per cent of the annual crop of flowers. The following figures are the approximate average yield per 1000 kg. of fresh orange blossom:

	Grams of oil
1. Distillation	from 900-1100
2. Volatile solvents[14]	550-600
3. Maceration	350-450
4. Enfleurage	about 100

[14] The yield of absolute is about 1000 grams.

Distillation has always been the principal process, but at one time it was applied for the *Orange Flower Water* rather than for the oil, which was not much valued, and looked upon as a comparatively insignificant by-product. As we all know today, things are very different and the oil fetches high prices. The aqueous distillate is, however, a much esteemed article of commerce, and its odour, like that of the *absolute, approximates* more nearly to the fresh blossom than does the oil. Immediately after distillation this is a pale yellowish liquid, which gradually darkens with age until it becomes quite red, especially if exposed to strong light for a lengthy period. Under these conditions the odour of the oil is impaired, and if its delightful fragrance is to be retained, it should be stored in airtight containers in a dark and cool atmosphere. There are a number of grades of neroli oil offered by all the Grasse houses, and while some of the differences in odour may be accounted for by the methods and time of distillation, yet in the lower grades the odour bears a distinct suggestion of the blending in of finest French petitgrain oil. In the course of distillation the water vapour undoubtedly acts upon the constituents of the oil, particularly upon the unstable terpenes. The esters also become partially saponified and the aldehydes more or less polymerised. Fortunately the alcohols, which constitute the most important part of the oil, are not seriously affected. Neroli oil remains one of the most esteemed raw materials of the perfumer.

An essential oil is extracted from orange flower water by means of acetone, petroleum ether, or ether, the yield being one kilo from about 3000 kilos of the water. It has a powerful odour, some ten times the strength of the normal oil, and owing to its solubility in water is useful for the extemporaneous production of Eau de Fleur d'Oranger.

Chemistry

The composition of neroli oil has been known since 1895, but the first complete analysis was not published until 1902. So far, the following constituents have been identified:

Terpenes, chiefly dipentene, pinene, and camphene	35 per cent
l-Linalol	30 per cent
Geraniol and nerol	4 per cent

Phenylethyl alcohol (found in the oil from the aqueous distillate)	traces
d-Terpineol	2 per cent
d-Nerolidol	6 per cent
Decylic aldehyde	traces
l-linalyl acetate	7 per cent
Neryl and geranyl acetates	4 per cent
Esters of phenylacetic acid and benzoic acid	traces
Indole	up to 0·1 per cent
Methyl anthranilate	0·6 per cent
Jasmone, farnesol, acetic and palmitic acid	traces

The characteristic odour of the oil is primarily due to methyl anthranilate (and possibly jasmone), although such a small quantity is present, but there appears to be very little doubt that it is also influenced by both indole and aldehyde C_1. The odour of some oils would suggest that geranyl formate might be a possible constituent, but so far there appears to be no analytical evidence in support of this.

Orange flower water

Orange flower water is a product of considerable commercial importance and in the majority of works is obtained as a by-product during the production of the oil. There are a number of distilleries, however, which attribute more importance to this product and these are situated at Le Cannet, Golfe-Juan, and Vallauris. Orange flower water of several strengths is recognised:

1. *L'eau 2 kilos.* 1000 kg of blossoms are distilled with a similar quantity of water and the first 500 litres collected.

2. *L'eau 1 kilo.* 1000 kg of flowers are distilled with 1500 litres of water and the first 1000 litres reserved. This is known as Eau quadruple or K.P.K.

3. *L'eau double* is a mixture in equal proportions of distilled water and No. 2. It is known also as L'eau codex.

4. *L'eau simple* is an equal dilution of No. 3.

5. *L'eau triple* comes between Nos. 2 and 3 and is often known as L'eau superieure.

6. *L'eau de brouts* is produced during the manufacture of oil of petitgrain during the summer months.

The sophistication of these several grades of waters has become a serious matter and led to an investigation by M. Bonis, who outlined certain methods of standardisation. This matter has also been studied by Tombarel Freres, who for the purposes of this investigation divided the water obtained in the course of distillation into two parts. For 100 kg of flowers they separated the first 80 litres from the last 20 in order that the weight of oil in the fractions might be assessed as distillation proceeded. The first 80 litres gave an average of 0·031 per cent of oil, whereas the last 20 litres gave an average of 0·008 per cent only. A mixture of these portions corresponding to K.P.K. gave an average of 0·027 per

Odour classification

Top notes	*Middle notes*	*Basic notes*
1. Benzyl acetate Methyl benzoate	21. Indole Methyl anthranilate Myrrh resin	77. Methyl naphthyl ketone
2. Linalol Mandarin Octyl acetate Rosewood	22. Geranyl formate Linalyl anthranilate Orange flower water absolute	79. Civet absolute 80. Hydroxy citronellal 100. Amyl cinnamic aldehyde
3. Decyl acetate Decyl formate Terpineol	24. Geranyl acetate Ylang	Decyl aldehyde Iso-eugenol
4. Citronellol Phenyl ethyl alcohol	26. Iso-butyl methyl anthranilate 31. Orange flower absolute	Methyl nonyl acetic aldehyde Musk ketone
5. Nonyl acetate	35. Dimethyl anthranilate	Phenyl acetic acid
6. Bergamot Citronellyl formate Iso-butyl phenylacetate Linalyl benzoate	43. Jasmin absolute 50. Laurinic alcohol Neroli	
7. Geraniol, Java	55. Nerolidol	
8. Ethyl anthranilate Iso-butyl benzoate Nerol Rhodinol	60. Benzyl phenylacetate Citral Rhodinyl formate	
9. Neryl acetate		
10. Linalyl acetate		
11. Decyl alcohol Linalyl cinnamate Sweet orange		
12. Methyl cinnamate Petitgrain, French		

cent of oil. These figures were confirmed by an examination of authenticated water from other sources. It should be noted that with age the per cent of oil decreases.

Orange flower assumes a greenish coloration with age and when exposed to strong light.

Petitgrain oil

Petitgrain oil is referred to as far back as 1694 by Pomet;[15] at that time it was distilled from the small unripe fruits about the size of a cherry which fall from the trees shortly after the flowers; they are called 'orangettes', and the origin of the name is attributed to these small seeds or kernels. Pomet says they were infused in water for five or six days before distillation. The orange trees are pruned about the end of June, and from these prunings are separated the leaves and twigs which on distillation yield the petitgrain oil of today. The finest oils are obtained from the bitter orange trees of France, although other oils are distilled in Tunis and Paraguay. The sweet orange, the lemon, and the mandarin also yield petitgrain oils of distinct type which are useful in perfumery, but in odour they do not approach neroli oil so nearly as the first-mentioned product.

Compounding Notes

Many of the finest neroli oil imitations are nothing more nor less than terpeneless petitgrain oil to which has been added in some cases neroli and aldehyde C_{10}: such products may be compounded as follows:

Neroli, no. 1048

800	Petitgrain oil, French, terpeneless
100	Nerol
10	Aldehyde C_{10}, 10 per cent
90	Neroli oil
1000	

Even when preparing less costly synthetic nerolis, French petitgrain is almost invariably employed, and in the cheapest of all products Paraguay petitgrain. The fragrance is improved by substantial additions of linalol, but less of this alcohol is necessary in the duplication of orange blossom. Nerol and/or its esters, as well

[15] *Histoire des Drogues*, 151.

as methyl anthranilate, are indispensable constituents of both oils, but there the likeness ends. For in neroli the following aromatics are especially useful: terpineol, indole, laurinic alcohol, hydroxy citronellal, amyl cinnamic aldehyde and phenyl acetic acid; whereas in orange blossom benzoates, phenylacetates and iso-eugenol are essential.

The condensation product of methyl anthranilate and hydroxycitronellal, known as aurantiol, is also a most valuable asset in all types of orange flower products.

Examples of each complex follow:

Neroli, no. 1049

250 Linalol
 80 Terpineol
 50 Nerol
 80 Neryl acetate
 70 Linalyl acetate
300 Petitgrain, French
  8 Indole, 10 per cent
 50 Methyl anthranilate
 10 Laurinic alcohol
 50 Hydroxy citronellal
 40 Amyl cinnamic aldehyde
  2 Decyl aldehyde, 10 per cent
 10 Phenyl acetic acid
─────
1000

Orange Blossom, no. 1050

 30 Benzyl acetate
 10 Methyl benzoate
100 Linalol
200 Phenyl ethyl alcohol
 20 Iso-butyl phenylacetate
100 Nerol
 40 Linalyl acetate
400 Petitgrain, French
 10 Methyl anthranilate
 30 Orange flower absolute
 50 Methyl naphthyl ketone
  1 Decyl aldehyde
  9 Iso-eugenol
─────
1000

A perfume of this type may be prepared thus:

Orange Blossom, no. 1051

100	Orange blossom, no. 1050
40	Neroli, no. 1049
5	Orange flower absolute
4	Jasmin absolute
1	Civet absolute
20	Ambergris tincture, 3 per cent
30	Musk tincture, 3 per cent
800	Alcohol
1000	

Orchids

History

The *orchis* is referred to in a few places in classical literature. Theophrastus mentions it, while Pliny discusses at length the medicinal properties of some five different varieties. In Book XXVI, chapter 62, this writer says there are few plants of so marvellous a nature as the orchis, a vegetable production ... having a twofold root formed of tuberosities which resemble the testes in appearance. The larger of these tuberosities, or as some say, the harder of the two, taken in water is provocative of lust; while the smaller, or, in other words, the softer one, taken in goat's milk, acts as an antaphrodisiac. Some commentators are of the opinion that these properties have not been proved, but Linnaeus, however, seems to think that the orchis may have the effect of an aphrodisiac upon cattle. It is the name, no doubt, signifying 'testicle', which originally procured for it the repute of being an aphrodisiac.[16]

The plant

Orchis is the name of a Linnaean genus and the type of an extensive family of plants termed *Orchidaceae*, generally referred to as orchids. The number of known species exceeds 6000, and they take the form of shrubs or herbaceous perennials, all of which are particularly handsome and deservedly popular. They are found in tropical and temperate countries; a large number grow on trees, some grow on rocks, and many have tuberous roots growing

---

[16] Consult also Pliny, Book XXVII, chap. 42, where a more detailed account of these properties is discussed.

in the ground, of which over forty are natives of this country. Although the species are numerous, the only one of economic importance to the chemist-perfumer is *vanilla*. In Asia, a farinaceous meal is obtained from the tubers of several terrestrial orchids. This is known as *Salep*, and in northern India it is largely used as food by the natives. In this very large group of plants, there is an unusually wide range of floral variations, those of form and colour being particularly noticeable. The flowers also exhale a variety of odours, ranging from the most exquisite fragrance to the most disagreeable putrescence. During recent years orchids have become highly patronised as peculiar and attractive garden plants, and to obtain them plant collectors have gone to the most remote regions of both hemispheres; they have thus become important trade plants. High prices have been obtained for rare varieties, and in 1881 an amateur collection realised over £5000, some individual plants being sold for as much as £100 or even more.

Odour varieties

A series of observations on the odours of orchids was conducted by E. Andre, an eminent French botanist, and among the varieties exhibiting peculiar scented attractions are the following:

Acropera Loddigesii, the dainty scent of wallflowers.
Angraecum fatuosum, a sweet odour of tuberoses.
Bifrenaria inodora, like muguet.
Catasetum scurra, the aroma of lemons.
Cattleya Eldorado, an odour of roses in the evening only.
Caeliopsis hyacinthesma, the perfume of hyacinths.
Cleistoma ionasmum, violet-scented.
Dendrobium glumaceum, lilac in the evening and heliotrope in the morning.
Denrobium nobile, odour of hay in the evening, honey at noon, and primrose in the morning.
Epidendrum vulnerum, an odour of carnations in the morning.
Macillaria aromatica, cinnamon-scented.
Odontoglossum citrosum, rose-scented.
Odontoglossum gloriosum, like a whole hedge of hawthorn.
Orchis Sambucina, the odour of elder flowers.
Zygopetalum Mackayi, hyacinth-scented during sunshine.

Two other odorous species are worthy of note:

Herminium monorchis, known also as the musk orchid, grows to a height of 4 inches, and during June and July bears attractive pale green flowers. Odour, musk-honey.

Haboenaria conopsea, known also as the sweet scented orchid, grows to a height of 4 to 8 inches, bears mauve flowers and has the perfume of hyacinths.

In China one of the most highly esteemed flowers belongs to the orchid family. This is *Cymbidium ensifolium*, and it exhales a very intense and sweet fragrance. Its cultivation requires considerable care—a very even temperature (about 15°C) and rain water only for watering.

Chemistry

From the point of view of the perfumer the most important orchid is vanilla, the cultivation, curing, and chemistry of which has already been discussed in Volume I and elsewhere. The most important constituent is vanillin together with anisic aldehyde, anisic acid, and alcohol. As far back as 1901, however, Crouzel extracted *orchis militaris L.* with ether or alcohol and obtained a small amount of oil, yellowish in colour and of a pleasant strong odour. The oil could not be obtained by steam distillation.

Compounding notes

It will have been observed from the foregoing that a wide variety of fragrances are exhaled by the scented orchids, and for this reason no single one of them can be taken as a standard by perfumers. Thus, compounds bearing this name are largely the creation of fancy, but at the same time it should be remembered that many of those on the market are dominated by salicylates. On the assumption that this is the most acceptable top note, the choice lies between amyl and iso-butyl salicylates, with a distinct preference for the latter owing to its more floral character. The synthesis might be started by taking a quantity equivalent to 50 per cent of the finished blend and then modifying it to the fragrance of some preferred flower. While this could be accomplished most easily by direct admixture with an already compounded flower oil, there are many perfumers who prefer to work with raw materials, and since the hyacinth note blends admirably with salicylates it is adopted as the background in the following

Odour classification

Top notes	Middle notes	Basic notes
1. Benzaldehyde 　 Benzyl acetate 2. Linalol 4. Nonyl aldehyde 　 Phenyl ethyl 　　　alcohol 6. Bergamot 7. Geraniol, Java 8. Amyl salicylate 　 Iso-butyl salicylate 　 Nerol 　 Rhodinol 10. Phenyl ethyl 　　　propionate 11. Sweet orange 14. Methyl Ionone	15. Cinnamyl acetate 　　 Heliotropin 21. Anisic aldehyde 　　 Ionone alpha 24. Ylang 31. Orange absolute 43. Jasmin absolute 　　 Rose absolute 　　 Tuberose absolute 50. Laurinic aldehyde	65. Cinnamic alcohol 79. Civet absolute 80. Hydroxy 　　　citronellal 100. Benzoin resin 　　 Castoreum 　　　absolute 　　 Coumarin 　　 Decyl aldehyde 　　 Iso-eugenol 　　 Methyl nonyl 　　　acetic aldehyde 　　 Musk ambrette 　　 Oakmoss 　　 Peru balsam 　　 Phenyl acetic acid 　　 Phenyl acetic 　　　aldehyde 　　 Vanillin

example, which is reduced to its vital constituents and may be elaborated to taste by choosing aromatics from the odour classification.

Orchid, no. 1052

60	Benzyl acetate
50	Bergamot
200	Geraniol, Java
500	Iso-butyl salicylate
50	Sweet orange
100	Methyl ionone
30	Phenyl acetic aldehyde
10	Vanillin
1000	

A finished perfume may be prepared from the foregoing as follows:

Orchid, no. 1053

140	Compound, no. 1052
5	Jasmin absolute

2	Rose absolute
1	Civet absolute
2	Tonka resin
50	Musk tincture, 3 per cent
800	Alcohol
1000	

Reseda

History

Mignonette is a native of Egypt and the shores of the Mediterranean. The plant was known to Pliny, and in those days the Romans used it as a charm to allay the irritation of wounds, disperse abscesses and all kinds of inflammation. The origin of the name *Reseda* may even be connected with its clinical value and is said to be derived from the Latin *resedo*, to heal. According to Pliny the plant grew near the city of Ariminium (now Rimini) in Italy. At an early date it was introduced to the south of France, where it received the name of mignonette, signifying in French 'little darling'. It made its appearance in Britain about 1750, and two years later is stated to have been cultivated in the Apothecaries' Botanic Gardens, Chelsea. Since then it has been a great favourite, and in this country is an annual, while in the south of Europe it becomes shrubby.

Varieties

Reseda odorata is the type of the N.O. Resedaceae, consisting of over twenty species. The flowers were originally greyish-white with red stamens, but have since given place to fine white, red, and yellow-flowered varieties.

    R. odorata has greenish-yellow or white flowers, highly odoriferous, more or less common in British gardens. This is the species cultivated in the south of France. There are several varieties of this species, the most important being the compact, strong-growing *Machet*. The flowers are reddish. This is much grown in pots for the London markets.

    R. luteola is a smaller plant having greenish-yellow flowers. It grows wild on waste chalky soil. This species is known in France as "Gaude" and *Reseda sauvage*. It is of no value as a raw material in the perfume industry, but is occasionally employed (and culti-

FLOWER PERFUMES: RESEDA

vated) as a yellow colouring material. In Britain this species is known as Dyer's Rochet and yellow weed.

R. lutea is a similar but less significant species.

R. alba is like *R. lutea*, but has whiter flowers.

Jamaica mignonette is a pseudonym applied in the West Indies to Henna flowers (*Lawsonia Inermis*). Their fragrance is more suggestive of mignonette combined with rose.

Odour

The perfume of the common mignonette grown in this country is poor compared to the very fine fragrance obtained by careful cultivation in the south of France. The odour recalls that of violet leaves and basil.

Natural perfume

In the south of France (Grasse, La Roquette, Mandelieu, and Pégomas districts) reseda is grown for its perfume, which is extracted principally by volatile solvents and to some extent by maceration and enfleurage. The plant is delicate and the crops often fail in consequence of late winds. The demand, however, for this flower by the factories at Grasse has noticeably declined, and since growers do not plant unless they are sure of selling the crop, its cultivation has diminished. It is usual for the growers to ascertain the requirements of the works during February, and plant accordingly, the crop occupying the ground only for five months from March to July. Mignonette requires not only a rich, fertile, clean soil with plenty of sun and irrigation, but it also needs ample manure and labour. There is always a risk of irregular yields. The flowers gathered in March and April produce the finest perfume. About 1200 kilos of flowers yield 1 kilo of concrete which gives some 350 grams of absolute. This is not very largely employed, but when violet crops are poor the demand for it is generally increased.

Chemistry

As far back as 1893 this plant drew the attention of research workers, and Schimmel & Co. were able to isolate about 0·002 per cent of volatile oil. This was a semi-solid substance having an intense floral odour. In 1895 Bertram and Walbaum distilled the root of this plant and obtained an essential oil having an odour recalling radishes. The principal constituent proved to be phenyl-ethyl thiocarbamide.

Mignonette perfume was for some years ingeniously marketed by Schimmel & Co. who distilled with every 500 kilos of blossoms 1 kilo of geraniol. This substance was called Reseda Geraniol and made a good base for the flower perfume.

Odour classification

Top notes	Middle notes	Basic notes
1. Benzyl acetate	15. Heliotropin	73. Cassie absolute
2. Linalol	16. Eugenol	79. Civet absolute
3. Ethyl decine carbonate	18. Violet leaf absolute	85. Phenyl acetaldehyde
4. Cummin	20. Clary sage	Dimethyl acetal
Phenyl ethyl alcohol	21. Anisic aldehyde	88. Ethyl methyl phenyl glycidate
	Ionone beta	
6. Bergamot	22. Clove	90. Galbanum resin
7. Geraniol, Java	24. Ylang	100. Benzoin resin
8. Cedarwood	43. Rose absolute	Costus
Iso-butyl salicylate	60. Citral	Coumarin
Lemon		Ethyl vanillin
Rhodinol		Iso-eugenol
12. Methyl heptine carbonate		Labdanum resin
		Musk ketone
Petitgrain, French		Orris resin
13. Orris concrete		Santal
14. Basilic		Styrax resin
Methyl ionone		Vanillin
Mimosa absolute		
Reseda absolute		

Compounding notes

The peculiar greenish-violet fragrance of mignonette will have been noticed by every perfumer, and since a number of aromatics of this type are available this particular tonality will present no difficulty in the creation of the complex. The ionones are therefore indispensable, together with one of the violet leaf synthetics where the choice lies between methyl heptine carbonate and ethyl decine carbonate with a slight preference for the latter; while a natural product that should not be overlooked is basilic in addition to violet leaf substitute. But these substances by no means solve this rather difficult synthesis; for if a large bunch of flowers is examined olfactically a peculiar fruity shading will be detected that perhaps suggests liberal quantities of bergamot and lemon, together with strawberry. Once this has been confirmed the build-up can be completed with natural reseda and mimosa, together with small quantities of the crystalline aromatics.

These suggestions are illustrated in the following formula:

Mignonette, no. 1054

20	Ethyl decine carbonate
100	Bergamot
100	Lemon
100	Rhodinol
20	Petitgrain, French
10	Orris concrete
3	Basilic
500	Methyl ionone
10	Mimosa absolute
20	Reseda absolute
30	Eugenol
30	Anisic aldehyde
2	Civet absolute
5	Ethyl methyl phenyl glycidate
30	Musk ketone
10	Coumarin
10	Vanillin
1000	

Since chypre blends admirably with reseda the following formula for a finished perfume is based upon this tonality:

Reseda, no. 1055

100	Mignonette Compound
40	Chypre Compound
5	Cassie absolute
1	Violet leaf absolute
4	Rose absolute
20	Ambergris tincture, 3 per cent
30	Musk tincture, 3 per cent
800	Alcohol
1000	

Rose

History

The charm and delightful fragrance of the rose is known and appreciated the world over today. Since earliest antiquity these qualities have led to its being regarded as the queen of flowers, and ancient literature contains many references to it.

It is doubtful whether several of these references are to the rose as we know it today, and this is well illustrated by one of the earliest mentions which occur in the Scriptures. It was written,[17] 'The desert shall rejoice, and blossom as the rose'. In this verse the Hebrew word has been translated rose while the same word has also been translated (300 years later) Rose of Sharon.[18] According to John Smith, one of the ablest commentators, it is very improbable that both plants are roses. Certain roses grow wild on Lebanon and in other parts of the Middle East, the most prolific being the Dog Rose, *Rosa canina*. The other references[19] to roses in the Apocrypha may, according to Smith, be any of the following: *Cistus ladaniferus, Hibiscus Syriacus, or Nerium Oleander*.

Mythological writers found ample scope in endeavouring to account for the creation of the rose. One of the oldest stories is that Flora found the corpse of a beautiful nymph, a daughter of the Dryads, and with the assistance of Venus and the Graces transformed it into this beautiful flower. The ceremony was attended by the Zephyrs who cleared the atmosphere so that Apollo might bless the new flower with his beams. Bacchus supplied the nectar and Vertumnus the perfume. Pomona strewed her fruit over the young branches, which were then crowned by Flora with a diadem especially prepared by the Celestials to distinguish this most beautiful blossom.

The perfume of roses was first extracted by placing the petals in oils and fats. They were then used as unguents, and Homer states that Aphrodite anointed the dead body of Hector with such a product.

Theophrastus tells us that the rose perfume was best absorbed by sesame oil because of its viscid nature. He also says that the rose imparts a fragrant scent or sweet taste to wines. There appears to have been some superstition associated with the collection of

---

[17] Isa. xxxv. 1. [18] Consult also Narcissus.
[19] 2 Esdras ii. 19; Wisdom ii. 8; Eccles. xxiv. 14.

the fruits of wild roses, for we read in his *Enquiry into Plants*, that if they were not gathered while standing to windward there was a danger to the eyes. Pliny describes the luxury in which the Romans lived, and says that their food was either covered with rose petals or sprinkled with rose perfume. In Book XXI, chapter 10, there is a discourse on several varieties of roses, and here Pliny remarks that flowers having 100 or more petals are known as *centifolia*. He further states, and quite erroneously, that the best proof of the perfume of the rose is the comparative roughness of the calyx. The most odoriferous of all roses, Pliny continues, comes from *Cyrenae*, and is the most esteemed for use in unguents. When the juice had been extracted from the petals they were dried and powdered. This particular *diapasmata* was sprinkled on the body to check perspiration.

Many considered the flower emblematic of joy while others regarded it as a symbol of silence. At feasts, where the conversation was to be held secret, it was customary to suspend a rose over the table. Hence the saying *Sub-Rosa*.

In Egypt the rose was equally popular, for we are told that men of rank had their mattresses filled with rose petals. Cleopatra was noted for her luxurious habits, and it is recorded that at one of the feasts she gave to Anthony, the royal apartments were covered with perfumed rose leaves to a considerable depth.

Roses evidently played an important part in the magic of ancient antiquity, for we learn on reading the metamorphoses (The Golden Ass) of the Roman jester, Lucius Apuleius, that while journeying through Northern Africa he committed some trifling indiscretion, and because of this was turned into an ass by a witch. The only way he could regain his human form was by *eating* roses.

The distillation of roses probably originated in Persia at a very remote period, which may quite conceivably ante-date the Christian era. According to a document in the National Library of Paris the province of Faristan was required to pay an annual tribute of 30,000 bottles of rose water to the Treasury of Baghdad as far back as A.D. 810. Faristan seems to have been the principal centre for the production of this commodity, for it is also stated that considerable quantities were from thence sent to India, China, Yemen, Egypt, etc. The most important factories appear to have been situated at Firazabad, between Shiraz and the coast. The art of distillation was probably introduced into western countries by the Arabs in the tenth century, and the first country in Europe to employ it would seem to have been Spain.

It is a curious thing that no mention is made of rose otto until 1574 when small drops of it were found floating on the surface of rose water by Geronimo Rossi at Ravenna. Its discovery by the Persians (1612) is associated with one of the Grand Moguls who filled the canals in his gardens with rose water. One of the Princesses noticed a scum floating on the surface which she caused to be collected. This was found to be intensely odorous and was highly treasured by her. The production of rose otto (*Aettr Gyl*) at Shiraz dates from this period.

About the seventeenth century rose cultivation spread from Persia to India, northern Africa, and Turkey, and in the year 1710 was established for the first time in Bulgaria, close to the village of Shipka (at the foot of the famous pass) in the Kazanlik valley. The Damask Rose plants were brought from Asia Minor and their cultivation thrived so well that in thirty years it extended to the whole valleys of Karlovo and Kazanlik. By 1750 Bulgaria had become the principal source of supply of the present rose otto.

In more recent years the plant was cultivated in England, France, and Germany, but the commercial distillation of the oil was not begun in France until the end of the nineteenth century. Previously the flowers had been grown for sale, and in some cases were used for the production of fragrant rose water. The large quantities of blossoms which were then wasted are now, as above stated, turned to some commercial account.

Since its first production, attar of roses has been much esteemed as a perfume, and many years ago the demand was generally greater than the supply. In consequence sophistication was much practised both by manufacturers and dealers, and among the articles used for this purpose were palmarosa and ginger-grass oils and spermaceti. Before the Second World War the decreased demand for rose otto, the keen competition for quality amongst distillers, and a greater knowledge and experience among buyers, led to the sale of oils of exquisite quality. Naturally these were more costly, and buyers who insisted on forcing down the price received oils analysing as 'pure' but which were never acceptable to an experienced nose. Nowadays the new régime in

Bulgaria has resulted in the disappearance of well known private brands, which have been replaced by communal oils.

Flower varieties

According to ancient mythology the rose was originally white in colour, and the red varieties were created when a rose-thorn pierced the foot of Venus, the blood escaping from the wound being supposed to have dyed the petals! At the time of Theophrastus it is certain that some different varieties existed, for he describes many physical differences and refers to the perfume. Pliny also distinguished them one from another by similar comparisons. Nowadays, on referring to horticultural catalogues, we find thousands of varieties of roses listed under as many different names. It appears probable that a large number of these are the fancy of the florist and have no scientific botanical classification, in some cases a grower having given another name to a hybrid which had previously been discovered by someone else.

Over two hundred species have been named, but it is doubtful if more than one-sixth of this number are specifically distinct. Of the roses which are natives of these islands about twenty are said to belong to England, four to Scotland, one to Ireland, and one to the Scilly Isles. According to the 'Hortus Kewensis' these are made to form seven distinct species, the most delightful wild rose being the sweetbriar or eglantine, *Rosa rubiginosa* or *R. eglantina*, whose fragrance is especially charming after a shower of rain.

The characteristic rose perfume is well developed in the following species:

Rosa centifolia, L., large pinkish or purplish-red flowers, fragrant, cultivated in British gardens and known as 'cabbage' rose. The 'Rose de Mai' of the French perfumer. Single flowered varieties grow wild in the eastern Caucasus.

Rosa Damascena, Mill., large red (and occasionally white) flowers, cultivated in Bulgaria. *Rose de L'Hay*, having large pinkish-white flowers and experimented with by French perfumers, is a cross between *R. Damascena, R. Jacqueminot*, and *R. rugosa*.

Rosa alba, L., flowers generally white but sometimes pale pink.

Rosa Gallica, L., crimson or deep red petals. In Britain a variety of this species is known as the 'Damask' rose. The 'Province' rose of France is also a variety of *R. Gallica*.

Odour

Slight differentiations in odour value are not perceptible to every one. The sense of taste is, however, generally more developed, and inconsiderable modifications of flavour are more frequently noticed. To the trained specialist, however, the merest graduation of odour is appreciable, and an expert florist will name the variety of rose *even in the dark*. The real rose odour is unique, and represents a *type* which is undefinable, incomparable, and at present inimitable by synthesis. This type is best represented by *Rosa Damascena*, cultivated in Bulgaria, and approximated to very closely by *Rosa centifolia*, cultivated in Provence. There are many variations from this type, and some of the roses which exhibit slightly different floral notes are enumerated below; they represent practically all those that the perfumer is required to imitate:

R. arvensis, also known as the Ayrshire rose—some varieties are myrrh-scented; *Banksian*, recalling violets; *Canina*, resembling mignonette; *Desprez*, fruity; *Eglantine*, whose leaves recall jasmin; *Macartnean*, apricots; *Marechal Niel*, the most delightful of the 'tea' class, having a somewhat fruity odour, resembling raspberry; *Moschata*, growing wild in Tunis and believed by some to be musky and by others to resemble pinks; *Muscosa*, moss-scented; *May*, recalling cinnamon; *Safrano*, recalling pinks; *Socrates*, resembling the peach; *Souveraine*, the melon; *Unique jaune*, the most charming of the 'noisette' type and having an odour of hyacinths. During 1929 a new rose, known as 'Portadown Fragrance', made its appearance. It is alleged to have taken eleven years to 'produce' and is the result of ingenious crossing by Samuel McGreedy & Sons. According to report the odour is a complex of tea rose and verbena. In the majority of cases red roses are more odoriferous than white ones, but a peculiarity possessed by a number of both kinds is that, when cut and placed in water, their fragrance appears to be more pronounced than when growing. Roses cultivated in a hot climate have a more powerful perfume, and it has been noticed that those flowers grown under glass develop a finer aroma than those which thrive in the open air.

Some roses, e.g. *R. gallica*, develop their perfume when dried, while others, e.g. *R. Damascena*, lose it under similar circumstances. Sawer states that before a storm the odour of a rose *seems* strangely increased, and suggests that this may be due to the oxidising influence of the ozone in the atmosphere or to the perceptive faculties being sharpened at such moments. Other peculiarities of the rose odour are (*a*) that no two flowers emit the same fragrance; (*b*) that different flowers from the *same* plant have never *exactly* the same perfume.

Natural perfume

The finest and most powerful form in which the rose perfume can be obtained is undoubtedly the *otto*, which is a product of distillation. Unfortunately, however, owing to the loss of phenylethyl alcohol, a large proportion of which remains dissolved in the rose water, the otto does not accurately represent the flower odour. This is more nearly approximated to by the 'absolute' obtained by means of volatile solvents, which method is largely employed in the south of France and Morocco, and to some extent also in Bulgaria and Isparta. There is a slight difference in the odour of these four absolutes, and a good deal of that produced in Bulgaria and Morocco finds its way to Grasse. The perfume may be extracted in the form of pomade by maceration, and in a very few instances were rose water is produced in Grasse, the oil is separated which contains the most highly odorous constituents and all the stearoptene.

Bulgarian rose otto

Constitutes the bulk of the world's production of this volatile oil. The flowers are cultivated in the famous Valley of Roses. This is situated about 200 kilometres east of Sofia, and extends along the southern slopes of the Balkans. It is accessible by road over the Klissoura Pass. The valley is about 160 kilometres long and varies in width from 5 to 30 kilometres. This includes ten districts, of which the most important are Kazanlik, Karlovo, Brezovo, Tchirpan, Nova Zagora and Stara Zagora. There are in fact two actual valleys—Toundja and Strema—with Kazanlik and Karlovo as their central towns. About 150 villages, with a population of over 200,000, are more or less engaged in the rose industry. They cultivate a large area of rose fields, yielding annually some millions of kilos of flowers. The most highly productive districts are from 300 to 700 metres above sea level.

The rose trees flourish best in sandy or light stony soil, since this does not so easily retain water like chalky soil and therefore reduces the danger of the young plants being destroyed by frosts during the winter months. The principal species cultivated is *Rosa Damascena*, Miller, and this accounts for nearly 90 per cent of the roses grown. It is not known in the wild state, but is a product of horticulture, the original plant being a hybrid between *R. Gallica* and *R. canina*, with the characteristics of the former predominating. The shrub attains a height of 4 to 6 feet and blossoms during the months of May and June. Six or seven blooms appear on each branch but in a good year there may be double this number. The perfume is most powerful when the flower opens. Both single and double varieties are grown. A small quantity of *R. alba, L.*, is also cultivated because it thrives better on poorer soil, is more hardy and yields nearly 40 per cent more blossoms. As a rule it is used mainly for hedging the fields of pink roses. The perfume is not so fine, it yields less otto and in consequence is not so valuable. Both single and double varieties are cultivated, the former yielding a higher percentage of oil. Even this, however, is only about one-third the yield from *R. Damascena.* In addition to these two species, small quantities of *R. centifolia* and *R. Stamboletz* are cultivated.

The rose trees are grown either from roots taken from overgrown fields or from cuttings. They are planted out in rows, generally in the autumn and sometimes in the early spring. Those transplanted during November take root more easily. The ground is kept free from weeds, ploughed and manured systematically. No blooms appear in the first year, few in the second, more in the third, and a maximum between the fourth and tenth years. After about ten seasons the shrub becomes exhausted and is rejuvenated by the removal of its branches. New shoots appear in the following year and the crop is again good in the second year. Rose trees thus carefully treated are productive of blossoms for twenty or thirty years.

The harvest takes place during May and the early part of June, but this depends on atmospheric conditions and the place of collection. According as the weather is dry and hot or cool and rainy during the season, the harvest may last from fifteen to thirty days. With the former conditions the time is shorter, the buds expand quickly and during the heat of the day lose some of their perfume with a diminishing yield of oil. Ideal conditions are gentle rain occasionally with a little sun.

The flowers are usually collected before they begin to open, and while covered with dew, a little in advance of sunrise. Those which are gathered later in the day emit a more powerful fragrance of less delicacy, and the resulting distillation product is not so sweet in consequence.

The blossoms are collected by young boys and girls and also by a number of old women. They place the flowers in small baskets and when full empty them into sacks, containing about 40 kilos. They are thus transported to the distilleries, sometimes a distance of 50 kilometres, when they are unloaded in a cool and shady place and gradually distilled during the day and the following night. It is customary to sprinkle the sacks with water, to keep them cool, to prevent undue evaporation of the perfume, and incidentally to prevent fermentation. There are four different systems of extracting the perfume from the flowers, as follows:

1. Open fire with small alembics.
2. Open fire with large alembics.
3. Steam stills including vacuum and rotary apparatus.
4. Volatile solvent extraction.

Up to 1902 all the rose otto exported from Bulgaria was distilled by the growers themselves in small tinned-copper alembics. At that time there were in all 2800 distillers, operating over 13,000 small alembics having a total capacity of nearly 15 million kilos of rose blossoms. Now the old system of distillation is being replaced by modern methods, involving the use of large improved stills. These modern factories distil nearly 80 per cent of the entire rose crop, and the growers only 20 per cent of it. The latter generally use the old-fashioned *gulapana*, which constitutes one of the out-buildings of the local farmer. The small alembics are some 40 to 43 inches high and have a base of about 32 inches diameter. There are two handles, one on each side for lifting when charging and discharging. They are constructed as near as possible to a stream so that the supply of water for distillation and condensation is assured. The apparatus is comparatively crude and consists of a truncated cone copper boiler having a capacity of about 120 litres. The helmet top is mushroom shaped and has a straight condensing tube fitted into one side. This runs through a wooden vat containing the cooling water, and where it emerges below there is a cavity in the ground which acts as a receptacle for the glass receiver. The stills are generally placed in rows on a low brick hearth. Wood is used as fuel.

On arrival at the gulapana the red and white roses are mixed and distilled at once. During the busiest times it is often necessary to spread the roses out in a cool place to await distillation when they are moistened with cold water to prevent fermentation. Ten kilos of petals and 75 litres of water are placed in each boiler, the apparatus is fitted together and sealed. The fire is then lighted and it takes from one and a half to two hours for complete distillation. The yield of rose water and otto measures about 10 litres and this is placed aside. The apparatus is then emptied, the exhausted roses separated by strainer and the hot water returned to the boiler for the second distillation. This generally measures about 50 litres, and so another 25 litres of fresh water is added for the further charge of 10 kilos of rose blossoms. This process is repeated until the day's collection of flowers has been disposed of. The rose otto is separated from the reserved distillates by placing the products of four distillations (40 litres) together in the apparatus and carefully distilling the whole. The first 5 litres of cloudy distillate are reserved, and when this clears the otto is found floating on the surface. It is carefully reserved and stored in glass flasks. The other 35 litres is left in the still and a fresh quantity of rose water added. The process is repeated and the otto separated.

As will be observed, the peasant-farmer generally reemploys the residuary water remaining after each operation, and this constitutes a fundamental difference between his process and that of the larger manufacturers, where the exhausted flowers and residuary water are discarded. In the former case, the hot and dark coloured residuary waters give the farmer the advantage that he commences his new distillation with already heated water and so saves fuel. It will be obvious that the progressive concentration of extractive matter in these residuary waters will raise the boiling-point and cause constituents to be carried over with the distillate which increases the yield of oil. Since, however, such residuary waters in time assume a sharp and distinctly unpleasant odour, the increased yield of oil is largely discounted by the fact that the quality is impaired. Nowadays many farmers have observed this disadvantage and, like the larger distillers, completely discard both exhausted blossoms and water together, thus dispensing with the use of sieve or strainer as the case may be. The rose oil made by this primitive method is known as Peasant Quality, and possesses a soft, sweet, honeyed odour lacking in strength. It is cheaper than the oil distilled in the large alembics because, in the first place, the farmer does not 'cost' his labour and that of his family, and in the second place the capital involved is comparatively insignificant. This type of oil successfully competes with ottos produced by other means of distillation.

There are a number of large distilleries now belonging to co-operatives. These contain large open-fire stills having a capacity varying from 500 to 2500 litres and are capable of distilling 100 to 500 kilos of roses at each operation. They are usually made of copper and may have either fixed or detachable heads. Different forms exist, some being the flat type of still and others the erect cylindrical apparatus. In all cases they differ from the small former type in that they are built into hearths so that no loss of heat occurs. Moreover, they have a perforated platform inside near the base to prevent the flowers coming into contact with the direct heat of the fires beneath. Worm condensers are almost invariably used, and the receivers resemble large florentine flasks having cylindrical glass tops for the observation of the oil as it collects on the surface. In the fixed-head type of still the flowers are charged through a manhole direct from the sacks and until the blossoms are within a foot of the top. Water is then run in until the flowers are completely covered.

A worker stirs the whole mass with a long pole to ensure the separation of the blossoms so that they will float freely once the operation has commenced. (Coagulated groups of flowers would interfere with the complete removal of the essence.) In normal times the ratio of water to flowers is strictly observed, being 4 or 5 litres to 1 kilo of roses. The ratio is only decreased with pressure of work when the crop is exceptionally abundant. The charging orifice is closed and the fire lighted. Considerable experience is here necessary. The application of heat must be gentle, otherwise a too sudden rise in temperature would drive out some of the lighter perfume constituents with the air remaining in the still and would thus be lost to the detriment of the subsequent oil distillate. As a rule it takes one and a half hours before any condensation takes place, and from two to two and a half hours more to complete the distillation. The distillate passes through the condenser, which is comparatively hot at the top and not very cold at the bottom. The temperature of the distillate is kept at just over 35°C, otherwise the stearoptene would crystallise inside the condensing tubes. The distillation is not stopped until about 1 litre of rose water is collected for each kilo of flowers in the still, the actual volume depending upon pressure of work. The contents of the still, that is, exhausted flowers and water, are emptied from a side exit and pass down shoots into a river. The distillate flows from the condensers into the receivers, arranged in series of two for each still, and the rose water constantly passes from the second into open tanks having a capacity of 100 to 150 litres. When full, the contents are automatically pumped into large storage tanks for subsequent redistillation or cohobation. The oil floats to the surface of the receivers and appears in the glass cylindrical top already referred to. It is usually a pale yellowish-green crystalline mass, containing all the stearoptene and the more highly aromatic constituents. This is the Direct Oil and is known locally as *Surovo maslo*. It is removed by either pipette or spoon and transferred to glass bottles for the time being. The yield is comparatively small, because the major portion of the oil remains dissolved or is emulsified and in suspension in the rose water pumped into the storage tanks. This is known as the first waters, and when a sufficient quantity has accumulated it is transferred to the stills for cohobation or redistillation. About 1200 litres are run into each still. The fire is lighted and heat applied gently. In an hour and a half the distillate begins to pass over and the operation is completed in another hour. The condensers are operated cold since no stearoptene is present in this distillate. When some 150 litres of rose water have been collected, the distillation is stopped and roses added to the residual waters in the still. (The difference between this residual water and the peasant residuary waters above referred to will be apparent.) Distillation commences again for the production of direct oil. The above 150 litres of rose water constitute the second waters and are pumped into the storage tanks to join other first waters therein and to await cohobation. The oil which separates in the florentine flasks from the distillation of the first waters is known as the Water Oil and is referred to locally as *Prevarka*. This is always fluid. It is removed and mixed with the Direct Oil to constitute the Rose Oil of commerce. The yield from the first and second distillation is, from the author's actual experience in Bulgaria, never a constant ratio; even though seemingly identical conditions were observed for repeated distillations. The approximate calculations made on the spot led to the estimate of 25 per cent direct oil as against 75 per cent water oil. These mixed runnings, as already stated, constitute the crude oil of roses. It is placed in glass bottles and exposed to the direct rays of the sun for a number of days, when impurities and colloidal matter are precipitated. The supernatant fluid is carefully decanted and filtered. Packing is done under the supervision of the local Excise authorities, who seal each of the well-known tinned copper vases.

Steam distillation

This is carried out by a few of the larger manufacturers who have very efficient and modern apparatus, often specially constructed after much expensive experimental work. The steam is usually generated in a building adjacent to the distillery and the stills are as a rule larger than those employed for direct fire, sometimes having a capacity of 3000 litres. Two types are in existence, one having a steam coil or steam jacket and the other having a perforated direct steam coil. The latter have the advantage of more quickly raising the temperature of the water in the still and incidentally replacing part of that distilled over during the operation of the plant. For all steam stills the process, flower ratios, etc., are much the same as those already described. Owing to the ease with which the steam can be controlled it is obvious that this constitutes a distinct advantage over the direct fire systems. It is not unusual for some of the larger manufacturers to have both steam and direct fire stills in operation together. The oils from the two sources differ in odour and the latter are usually somewhat stronger and sharper, even though the bouquet is less fine. From the author's comparisons on the spot, a mixture of the two would seem to make the ideal rose otto, blended in the ratio of 2 of fire to 1 of steam oil. The largest plant has over 70 fire stills, 2 enormous steam stills and 4 concrete batteries, all of which were in continuous operation a few years ago during the peak period of perfumery sales.

Vacuum distillation

There is one large plant in operation and the process differs slightly from the foregoing. In the first distillation no oil is separated, the product being entirely rose water which is pumped into storage tanks. It is then redistilled, and yields at one operation the whole of the oil. The odour of this rose otto is of a very special character and in the author's view would be exceptionally suitable for perfumes of the white rose type.

Rotating apparatus

This is a modification of the plant used in Grasse for the extraction of flowers with volatile solvents, and was introduced into Bulgaria by Charles Garnier many years ago. It consists of a fixed vertical drum, inside of which perforated metal drums containing the blossoms revolve on a horizontal axle. They continually dip into the boiling water at the bottom of the apparatus and the flowers are rapidly exhausted. The actual period of extraction is considerably reduced, roughly to half that required for the fixed apparatus. Moreover, the ratio of water to flowers is less, being somewhere between three to two and two to one. Furthermore, the quantity of rose water distilled over is only about half of that necessary in the fixed apparatus. Whereas in the latter the ratio of water oil to direct oil is about three to one, it is about two to one in the rotating apparatus. The principal difference in the oil obtained is that of a much lower stearoptene content with a consequent increase in the odour to weight ratio.

Volatile solvent extraction

A few of the modern factories employ this process, which was first introduced into Bulgaria in 1903 by Charles Garnier. The process is the well-known one, described earlier in this work. Curiously enough, as already stated, much of the rose concrete finds a wider application in Grasse than elsewhere.

Yield of rose otto

There are approximately 750 rose flowers to 1 kilo, and since about 3500 kilos of blossoms yield 1 kilo of oil, it takes 170 Damask roses to produce one drop of attar of roses. However, these figures are subject to wide variation, from 3000 kilos or less when a good yield is obtained to 4000 kilos and upwards for a poor one. The causes of these variations are to be found mainly in the atmospheric conditions. For instance, when the weather is intensely hot and dry, and especially when high winds are prevalent, the yield falls to its lowest owing to evaporation of the essence from the flowers before they are picked. Alternatively during mild conditions with occasional rain, the yield increases tremendously. The variations in the yield from steam and direct fire stills is not very great, but the advantage is generally with the latter. This may be accounted for by the higher temperatures dissolving out certain salts from the flowers, which raises the boiling-point in the apparatus with the consequent distilling over of heavier fractions of the essence. In the case of the peasant alembics the yield is even greater, and figures as low as 2500 kilos of flowers for 1 kilo of oil have been observed. This seems to be explained by the use of residuary waters previously referred to and also to prolonged distillation. While this may seem advantageous to the farmer, the forcing of the process doubtless carries over inodorous matter, which in the larger apparatus remains in the flowers. These portions act more as a natural diluent and are of no interest when the product is considered strictly from an odour standard. No data is at present available as to the constitutions of these bodies. There is a distinct and marked difference between the odour of this peasant oil and the others. The latter are characterised by a much more powerful and sharper bouquet—the strength of which is apparent after exposure on strips of clean absorbent paper. On the other hand, the farmer oil is readily distinguishable by a sweeter, more honey-like bouquet of considerably less strength. Rose otto from the large stills, having a stronger and sharper odour, was at first less appreciated by perfumers, who preferred the soft sweetness of the small distillers. In order to meet this taste, large distillers were compelled to blend their own distillate with farmer oil. All leading perfumers now prefer the strong, sharper oil because they get the same result by dilution when necessary.

Rose trade

Years ago this was done by buying the otto from the peasant distiller, when the whole risk was borne by him. It was the practice for the exporter to buy only when he needed the oil, or as was more frequently the case, when he had already sold it. Up to the year 1933, with the evolution of the large distiller, the whole business became more complex. It was necessary to have agents in the villages to arrange for the purchase of flowers for the coming season. The distiller had to act as a banker and advance money to the cultivator for his flowers. He had to distil himself and also to export. Moreover, following the exchanges of the different countries to which he sent his otto was imperative, if he did not wish to lose heavily on these transactions. In 1933, for various divergent reasons, the banks and co-operatives assumed responsibility for the whole rose industry and acted in all the above capacities. This does not appear to have proved a successful departure because the 1935 crop was distilled by the private distiller, presumably with certain restrictions as to output, unrestricted production apparently being resumed during 1936.

Moroccan rose otto

The Moroccan rose industry was originally established before the Second World War, but has since been rapidly developing; so much so that it is now comparable with those of Bulgaria, Anatolia and France. Production is centred in two widely separated areas: in the *Dades Valley* the principal centre is at El Kelaa des Mgouna, 300 kilometres south-east from Marrakesh on the southern slopes of the Atlas Mountains and adjacent to the Sahara Desert, at an altitude ranging between 3000 and 5000 feet. The area was formerly very wild, but has now been opened up owing to exploitation of its mineral resources, especially manganese. The roses occur as boundary hedges between small irrigable fields which flank the Dades River near its source. Since the inhabitants are dependent upon the grain production of the fields for their food supply the quantity of rose has been limited, but has nevertheless increased enormously since European exploitation expanded some twenty years ago. The variety differs from *Damascena* in the number of petals and general aspect, and the plants were probably brought there from Persia by the Arabs ten or twelve centuries ago. Two factories have been built to treat the rose flowers by distillation and extraction with volatile solvents. These are the property of UOP Fragrances and of CAPP. Some 2000 tons of roses are treated in a good year, though the crop may be seriously affected by frost and in 1952 was even totally destroyed. The harvest normally occurs during the last two weeks of April, and owing to the rapid advent of hot summer conditions rarely lasts beyond the end of May. The yield of otto is also very variable—from 3000 to 5000 kilos of flowers only 1 kilo of oil. The greater part of the crop is treated by distillation and production may amount in a normal year to about 450 kg. of otto, plus nearly 3000 kilos of concrete. The otto is of good quality but lacks the strength of the Bulgarian product. Roses in the Dades Valley are entirely free from parasitic diseases, which may be due to the cold winters and very dry summers. In Northern Morocco Chauvet has developed very large fields of roses at Khemisset, Tedders and Maaziz, the total area under cultivation attaining several hundred hectares. These are among the largest flower producing fields in the world, and when the blooms are at their best they present a spectacular appearance. The crop of rose blossoms exceeds 1000 tons and may well increase still further. Rainfall in this part of Morocco is sufficient for cultivation, and, therefore, irrigation is unnecessary. Chauvet has chosen to grow Rose de Mai and in consequence his product differs in bouquet from that produced in the Dades Valley.

Moroccan rose oils are welcomed by the perfumer owing to the present prohibitive cost of the Bulgarian otto, and although they have not quite the intensity of sweetness that characterises the latter, they may nevertheless be widely used in perfumes, either alone or in partial admixture according to the cost price required by the formula. For a detailed account of this industry, consult 'Roses of the Dades' by Georges Igolen. SPC, December 1970, pages 773 to 785.

Anatolian rose otto

Anatolian rose otto was formerly not a serious competitor with the foregoing. The industry was established about 1894 by a Turkish rose farmer who smuggled the rose trees into the country from Kazanlik. The chief centre of the industry is at present situated in the provinces of Isparta and Burdur. On behalf of Yardley these centres have been visited on two occasions by J. Hackforth Jones, who writes as follows:

'Although the actual production of Turkish rose otto has not greatly increased since before the Second World War, its relative importance is now much greater owing to the reduced production in Bulgaria. It has always been hard to account for the difference in odour between the two sources of peasant otto; and the only suggestion that can be offered is that distillation is certainly continued for a longer time in Turkey than that current in Bulgaria.

'The Turkish rose appears to be true *Rosa Damascena*, pink flowers with 25 to 30 petals. Cultivation methods are identical with those in Bulgaria but yields seldom exceed 1 : 3500, which suggests that the Turkish soil or climate is less suitable than the Bulgarian. The severe continental climate of Anatolia may be in part to blame, as also may be the exceptional altitude at which the roses are grown—the rose-producing provinces lie at an altitude of more than 3000 feet.

'Interest has focused on the newly constructed factories for the production of rose otto and concrete. The Rose Co-operative at Isparta has completed two plants with capacities respectively of 25 kg per annum rose otto and 250 kg per annum rose concrete. Another Co-operative, known as Gulbirlik, has built a much larger rose distillery at Islamkoyu. The Turkish government financed both projects and it is clear that they attach great importance to the modernisation of the industry.' More recently the Grasse firm of Robertet has established a distillery producing an excellent oil.

An olfactic examination by the author of the otto distilled by Gulbirlik reveals a marked improvement in bouquet, with the complete elimination of the background odour reminiscent of 'boiled cabbage' which characterises the average sample of communal oil from Anatolia.

French rose otto

French rose otto is produced on a very small scale, and the quantities offer no serious competition with Bulgaria. In general the plant is cultivated largely for the blossoms, which are extracted mainly by volatile solvents and by maceration. The principal centres are: La Colle, Grasse, St. Paul, Vence, Pegomas, Tourettes, Montauroux, Le Vignal, Opio and Mougins.

The communes of La Colle and St. Paul are the two principal centres of rose cultivation, not so much because of the nature of the soil, as undoubtedly because the insufficiency of water supply available for the work of irrigation does not allow the cultivation of other crops more remunerative.

The principal species cultivated is *Rosa centifolia, L.*, known in France as the Rose de Mai, of which there are two or three varieties. Other varieties, such as Louis van Houtte and Marie van Houtte, were distilled, and today probably enter the composition of oils of mixed origin. Other varieties of garden roses are now worked up for their perfume since they cost considerably less than the Rose de Mai, but yield concrete and absolute having a delicious rose odour.

The cultivation of roses requires considerable care, if good results are desired, and especially so when virgin soil is used for the planting. The land is first lightly ploughed in September and again to a depth of about 2 feet in October—motor ploughs being employed. The fertility of the soil is enhanced by the application of fertilisers and manure. January is the best month for planting, but when large numbers of shrubs are placed it is better to commence the work in November so that it can be completed by February. The most satisfactory results are obtained with plants prepared a year ahead in a nursery. These are placed in rows 1·40 metres apart with a distance of 50 cm between each plant. This requires about 10,000 shrubs per hectare. Good irrigation is necessary, but in the first year the yield of blossoms is poor. In the autumn the trees are pruned and the ground well manured. In the following May the first real crop is collected. This reaches its maximum in the fourth or fifth year and maintains its yield until the tenth year, when it begins to decline. The crop varies from 2000 to 5000 kilos per hectare and is influenced by the situation, atmospheric conditions, and the variety of plant cultivated. The gathering is done by women and children in the early morning and while the dew is still on the blossom. The perfume is then at its best and they are paid for their work by the weight of petals collected. These are transported to the factories as quickly as possible and spread out on the floors before treatment. Approximately 70 per cent are treated by volatile solvents, 20 per cent by maceration, and the remainder by distillation. In the former case petroleum ether is the principal solvent used, but when benzole is employed the yield is increased. The odour, however, is less fine and lacks strength. This product is used for cutting the other or can, of course, be purchased separately when desired and specified.

Small quantities of rose otto are distilled in Germany, India,

Tunis, and Persia. In the latter country the principal seat of the rose-water industry is at Maimand in the province of Fars, some hundred miles SSW of Shiraz. The oil is only separated when specially ordered. In Italy Rose de Mai is cultivated principally between Ventimiglia and Andorra. From 6000 to 10,000 kilos of flowers yield 1 kilo of rose oil.

Generally in commerce rose otto is distinguished according to its geographical source, Bulgarian, Anatolian, Moroccan, French, etc., and also according to its make. In 1968 the crops were approximately as follows (in tons): Bulgaria 5000, U.S.S.R. 5000, Morocco 2500, Turkey 2000, France 300. One kilo of oil is obtained by distillation from the following weight (approximate) of flowers:

		kilos
1.	Bulgaria	3,500
2.	Anatolia	4,000
3.	Morocco	3,500
4.	France	12,000

The low yield obtained in the south of France is due to the fact that frequently the blossoms are distilled primarily for *rose water* and the oil considered a by-product.

Through the courtesy of Roure-Bertrand Fils, the author was able to make some experimental distillations at their works in Grasse. Similar conditions were observed to those operating in Bulgaria, both direct and water oils being obtained. The following data emerged from their examination (figures per cent):

	Total Alcohols	*Citronellol*	*Stearoptene*
Direct oil	9·0	5·5	71·3
Water oil	91·9	75·8	traces only
Total oil	28·4	17·3	59·8

On the basis of these experiments, it requires 12,000 kilos of flowers to produce 1 kilo of direct oil, plus 9,000 litres of rose water. Similar experiments at the works of Lautier Fils gave a direct oil as follows:

	per cent
Total alcohols	12·2
Citronellol	6·3
Stearoptene	76·7

Chemistry

The following constituents have been observed by different chemists in commercial samples of rose oil:

	per cent
l-Citronellol (principal constituent)	24 to 64
Total alcohols as geraniol	63 to 84
Nerol	up to 10
Phenylethyl alcohol[20]	up to 1
Esters of geraniol	up to 3·5
Eugenol	up to 1
Stearoptene	7 to 25
l-Linalol	
Farnesol	traces
Citral	
Nonyl aldehyde	

The recorded chemical and physical constants for commercial rose oils are:

Specific gravity at 30°C	0·849 to 0·865
Rotation°	− 1 to − 5
Refractive index at 25°	1·452 to 1·466
Melting-point °C	15° to 24°
Alcohols as geraniol, per cent	63 to 84
Citronellol/Rhodinol, per cent	24 to 64
Stearoptene, per cent	7 to 25
Acid value	1·5 to 3·8
Ester value	3·7 to 17·5

The above figures include rose otto made by rotating apparatus, which as previously stated yields oils having a very low stearoptene content.

The most notable difference between these constants and those given in the early editions of this work is the high citronellol content admissible for Pure Rose Ottos. This drastic change actually dates from the author's visit to Bulgaria in 1932, when he personally controlled the distillation, in large fire stills, of a considerable purchase of oil by the firm of Yardley & Co. Ltd. This oil, which was of undoubted purity, contained 55 per cent of citronellol, and it quite naturally disturbed the opinions of analysts who in England had been accustomed to reject samples containing more than 35 per cent or thereabouts of this alcohol.

[20] Oils by vacuum distillation may contain higher proportions.

As a proof of this, in 1932, E. J. Parry stated that the maximum citronellol content of pure Bulgarian rose oils was 35 per cent, and that during the preceding eight to ten years he had met with samples of doubtful purity containing from 40 to 55 per cent of citronellol. This opinion was supported by the standard works of Gildmeister and Hoffman who placed the maximum figure as 37 per cent. There were, however, prior to the above date, published statements by A. Chiris, who in 1924 indicated for one sample 48 per cent, in 1929 for another sample 49 per cent, and in 1930 for a large number of pure samples, up to 55 per cent citronellol. Such a radical change as that provoked by the author's statement naturally led to observations and further investigations by other chemists. Charles and Robert Garnier stated that normal oils, distilled by themselves in Bulgaria for some years past in modern stills, averaged from 45 to 55 per cent citronellol. In cases where they had used rotating apparatus, this figure went up to as much as 63 per cent.

E. J. Parry and J. H. Seager made an exhaustive examination of authenticated samples of rose oil distilled by the co-operatives during the 1932 season and representing more than 75 per cent of the total crop. They found 45 to 61 per cent of citronellol in these oils, which figures were much higher than the limits previously cited by one of them.

In commenting upon this investigation L. S. Glichitch and Y. R. Naves expressed the opinion that such results were by no means new, but had been observed by them in numerous samples of undoubted purity since 1924. Robert Garnier confirmed this statement and observed that in his opinion all good Bulgarian rose ottos, distilled in large alembics and not containing too high a proportion of oil from white roses, contain more than 40 per cent of citronellol and generally from 45 to 60 per cent. This author in collaboration with Sebastian Sabetay reported on the characters of samples of 1933 crop oil when they found from 46 to 61 per cent of citronellol in rose otto obtained from different types of stills and districts. They further confirmed that ethyl alcohol is a natural constituent of the oil, a view which is not universally accepted.

E. J. Parry and J. H. Seager noticed a lower tendency in the citronellol percentage of the 1933 crop oils. Fifteen samples distilled by the co-operatives gave an average of 44·6 per cent. A remarkable point emerged from their figures. The highest citronellol content (51·1 per cent) was in an oil distilled at Pavel-Bania, whereas Garnier and Sabetay in their above report observed 60·9 per cent—a striking difference of nearly 10 per cent in the same year oil from the same district.

From these observations, all by chemists of repute whose only interest is in the establishment of the true standards for pure oils, it is obvious that the old figures must be amended. The indications at present are that the best rose otto, that is oil from the large stills, should have a citronellol content of from 45 to 55 per cent and that any sample below 40 per cent should be regarded with suspicion. It seems very doubtful whether any explanation will ever be found even for the wide differences observed for acknowledged pure oils. There is no doubt that the atmospheric conditions, the soil and district, the time of collection of the blossoms, the time elapsing before their distillation with the possible effects of fermentation, the ratio of red to white flowers, the type of still and also the *modus operandi* of the distiller, do affect the composition of the oil, to say nothing of its odour. As a proof of this, during the 1932 crop the co-operatives distilled at numerous villages, of which two will serve to illustrate this point. One is Sopot, situated on the north side of the Valley of Roses, towards its western end, and the other Derelie, some 15 kilometres distant on the south side of the valley. Both villages drew their flowers from the intervening district at the same time and used the same methods of distillation. Yet the oil from Sopot only contained 45 per cent citronellol as against 58 per cent in the Derelie oil.

As to the differences between the citronellol content in the previous ten years compared with that of former years, the author is driven to one of two conclusions:

1. That the evolution of the larger alembics has been responsible for the change, coupled with the fact that distillers have been compelled by keener competition to sell comparatively pure own distilled oils, their own distillation having in many cases formerly been blended with peasant oils where the process had been pushed to its extreme limits.

2. That rose otto has been deliberately and systematically adulterated with geraniol and recovered or synthetic stearoptene in quite large proportions for many years, and that previously published figures have been based upon this sophisticated oil.

In the event of the latter supposition subsequently being confirmed, buyers of rose oil would be led to some very disquieting conclusions. The implication is obvious, for it would mean that the analyst has been for many years rejecting the best and purest oils on account of their high citronellol content, and accepting as pure the very rose otto which had been adulterated.

Even today it is a comparatively easy matter to sophisticate rose otto scientifically because of the facility with which an artificial rose oil may be compounded from the known constituents. Some years ago the author made such an oil and published its analysis by J. H. Seager. The figures came within the usually accepted limits as follows:

Specific gravity at 30°	0·8547
Optical rotatum	−1·7
Refractive index at 25°	1·4624
Congealing point	21·8°
Acid value	1·9
Ester value	11·6
E.V. after acetylation[21]	212·3
Citronellol by formylation	31·2 per cent
Stearoptenes	18·0 per cent

Some further experiments in this connection will be referred to later.

It has been known for many years that the phenylethyl alcohol content of roses was lost during the distillation process, but it is only recently that any practical examination of the problem has been made. E. S. Guenther and R. Garnier suspected that this loss occurred when the exhausted flowers and residuary waters were discarded after distillation. They cooled and filtered these residuary waters and extracted them with volatile solvents. The resulting product was examined by H. Walbaum and A. Rosenthal, who found it to consist mainly of phenylethyl alcohol together with some eugenol.

Evaluation of rose otto

To the experienced nose, the odour of rose otto is subject to considerable variations, and these are probably due to a variety of circumstances. There are two well-defined types:

21 This equals total alcohols as geraniol—69·5 per cent.

1. The soft honeyed sweetness of the peasant oil.
2. The intense sharpness of the essence from the large fire stills.

These two comprise the main part of the oils distilled, but in addition there are:

3. The aldehyde character of the steam distilled oils.
4. The leafy heaviness of the product from vacuum stills.

There are also variations in the odour of the first two types, and the factors which may be responsible for these dissimilarities are:

(*a*) The ratio of red to white flowers.
(*b*) The time which elapses between picking and distillation.
(*c*) Fermentation when supply exceeds still capacity or alternatively the time taken in transit to the distillery.
(*d*) The particular villages from which each distiller buys his flowers to make his own special bouquet.
(*e*) The possibility in large fire stills of slight burning of the blossoms, due to contact with the walls of the apparatus.
(*f*) The actual type of fire still, some preferring the detachable and others the fixed head.

With regard to (*d*), some of the larger firms who have several plants acknowledge the fact that the oils obtained during the same season from these different districts have a dissimilar and yet characteristic odour. Naturally these houses make a point of blending the oils from different sources so that deliveries are uniform, unless, of course, one of the large buyers states a preference for a particular oil.

Now the difficulty which faces important buyers is to know the type of oil to purchase and the source to choose. Confidence in the distiller is imperative. It is true that when the co-operatives assumed control of the rose industry the source of supply was narrowed down, the selection being made from the alternative samples submitted. However, this régime has now ceased, and the author therefore proposes to discuss this question impartially and irrespective of the producer and his apparatus. In the past, buyers have quite naturally had some anxiety as to the purity of the rose otto for which they were paying a high price, owing to the reputation for sophistication, regrettably but perhaps not wholly unwarrantably associated with this product.

The fact that the constituents are known approximately and the chemical and physical constants fall within certain prescribed limits has, in the past, induced some buyers to resort to the opinion of the analyst. But since chemistry has come to the aid of the sophisticator, the reliability of chemical analysis is no longer acknowledged, excepting perhaps as an indication of the crudest adulteration. It must be borne in mind that the prime characteristic of rose otto is the fineness of its odour, and unfortunately buyers are left only with an olfactic assessment, and even this in the most experienced is subject to certain vagaries.

Some years ago the author had occasion to discuss this problem with several of the leading perfumery houses in Paris, Grasse, Switzerland, and Germany. The opinions of their technical directors, chief chemists and perfumers were solicited. The system employed by all of them was the same. The best sample of the present year was taken as the odour standard for the succeeding year, bearing in mind the comparative crudity of freshly-distilled oil. The favoured sample was in one case only subjected to analytical examination, the constants being used as an additional check on bulk deliveries.

It may be that some readers will have doubts as to the soundness of this procedure, but the author recently made the following experiments in collaboration with a well-known essential oil analyst, and the results will indicate the unreliability of a chemical examination.

A quantity of artificial rose otto was prepared more or less on the lines of the known constituents, but containing 0·3 per cent of aldehydes and 15 per cent of synthetic stearoptene. It was matured for one month to develop roundness and intensity of odour. Eight samples of rose otto were sent to the author's collaborator for examination, as follows:

1. 80 per cent fire distilled oil + 20 per cent palmarosa geraniol.
2. 80 per cent steam distilled oil + 20 per cent citronellol.
3. 30 per cent steam distilled oil + 70 per cent fire distilled oil.
4. 80 per cent fire distilled oil + 20 per cent rose compound.
5. 50 per cent fire distilled oil + 30 per cent steam distilled oil + 20 per cent rose compound.
6. 100 per cent fire distilled oil.
7. 90 per cent fire distilled oil + 5 per cent geraniol + 5 per cent synthetic stearoptene.
8. 93 per cent fire distilled oil + 7 per cent peasant oil from white roses only.

The results were not surprising. Samples 1 and 2 (crude

adulteration) and 8 were rejected and the remainder (including 4 and 5 skilful and 7 semi-crude sophistication) were passed as pure oils of good quality. This evidence would seem to be conclusive proof of the case against an assessment on an analytical examination only.

Summing up, therefore, the two alternative methods of purchase of rose otto, and having in mind many years' experience of the product, the author is of the carefully considered opinion that the ratio of reliability is about 75 per cent olfactic as against 25 per cent chemical evaluation by so far known methods. On reflection, this would seem to be the logical conclusion, because the consumers of rose otto sell odour and must therefore buy on this basis.

Olfactic examination

All essential oils have two well-defined attributes—Bouquet and Strength. They may be assessed separately, the results being ultimately collated. An examination of the bouquet is made in the case of rose otto by adding one drop of the oil to 100 cc of distilled water at 42°C in clean brandy glasses. Each glass may be examined carefully without olfactory fatigue, and all of them placed in the order of preference. This process is repeated every hour for six hours and then on the following morning. A note of the order is made after each examination and it will generally be found that certain samples persistently fall into the first places. An assessment of the strength is made by impregnating clean strips of absorbent paper with a standard quantity, usually in the case of rose otto 0·05 gram, and exposing them at room temperature under identical conditions. The odour is observed and noted between 10 and 11 a.m. daily for ten days. The results are classified in the order: strong, 3 points; medium, 2 points; and weak, 1 point. After the tenth day the points are totalled, and if those having the highest figures correspond with those taking first places under the bouquet tests, then the best quality product is indisputably established.

In order to check the value of this method of odour evaluation the author adulterated an authenticated pure rose oil with 5, 10 and 20 per cent of a rose compound analysing as pure rose otto. The *strength tests* were the only means of detecting the adulteration, the weakening of the odour of the oil being in ratio to the

Odour classification

Top notes	Middle notes	Basic notes
1. Almonds	15. Cinnamyl acetate	65. Cinnamic alcohol
Benzyl acetate	Guaiac wood	79. Civet absolute
Iso-butyl acetate	Heliotropin	80. Hydroxy citronellal
2. Linalol	16. Eugenol	85. Phenyl acetaldehyde
Phenyl ethyl acetate	17. Tetrahydro geraniol	Dimethyl acetal
Phenyl ethyl propionate	18. Orris absolute	88. Ethyl methyl phenyl glycidate
3. Benzyl cinnamate	Phenoxy ethyl iso-butyrate	90. Rhodinyl acetate
Dimethyl benzyl carbinol	19. Phenyl ethyl iso-butyrate	91. Phenyl ethyl phenylacetate
Terpineol	Verbena	Undecalactone
4. Citronellol	21. Ionone alpha	100. Acet eugenol
Dimethyl octanol	22. Benzyl iso-eugenol	Ambergris
Geranyl benzoate	Clove	Benzoin resin
Nonyl aldehyde	24. Ethyl cinnamate	Benzophenone
Phenyl ethyl alcohol	Geranium Grasse	Coumarin
5. Ethyl phenylacetate	Geranyl acetate	Decyl aldehyde
Nonyl acetate	Rose otto	Guaiyl esters
6. Bergamot	Ylang	Labdanum resin
Citronellyl formate	27. Citronellyl benzoate	Musk ketone
Iso-butyl phenylacetate	30. Ambrette seed	Patchouli
Phenyl ethyl benzoate	40. Rhodinol propionate	Phenyl acetic acid
7. Geraniol, Java	41. Mace	Phenyl acetic aldehyde
Methyl octine carbonate	43. Jasmin absolute	Santal
Nonyl alcohol	Rose absolute	Styrax resin
Geranyl iso-butyrate	45. Cassia	Tolu balsam
8. Benzyl salicylate	50. Laurinic aldehyde	Tonka resin
Citronellyl acetate	Undecylic alcohol	Trichlor phenyl methyl carbinyl acetate
Geraniol palmarosa	Neroli	
Lemon	54. Cedryl acetate	Undecylenic aldehyde
Nerol	60. Citral	Vanillin
Rhodinol	Rhodinyl formate	Vetivert
9. Dimethyl nonenol		
Geranyl butyrate		
Neryl acetate		
10. Diphenyl methane		
Diphenyl oxide		

(col. cont. on p. 193)

Top notes—cont.
11. Decyl alcohol
    Nutmeg
    Octyl alcohol
12. Methyl cinnamate
13. Orris concrete
    Phenyl propyl alcohol
14. Basilic
    Methyl ionone
    Palmarosa

proportion of adulterant used. For instance, the 20 per cent sample was the first to show a marked difference in odour at the end of four days and the 10 per cent at 7 days. In the 5 per cent, the difference was undetectable with certainty. The *bouquet* tests can only be taken as valuable in conjunction with the *strength* tests. That is to say, if an oil has a fine bouquet but is weak, then it would be suspicious, whereas a strong oil of poor bouquet would indicate that the stills, the method of operation, or the flowers in that particular instance were capable of improvement. If, however, as already stated, the same oil appeared high in both tests, then it could reasonably be assumed to be of the finest quality.

Compounding notes

Artificial rose oils are in constant use as components of a wide range of flower and fancy perfumes, and several of them are to be found in every laboratory. The most important member of this fraternity will in all probability be a synthetic rose otto, compounded with or without the natural oil, because it is useful not only as a diluent of the expensive natural product, but also as a flowering agent in other bouquets. Among the other rose compounds are sure to be those of red and white rose which, however, have a more specialised application. The paramount importance and extensive use of these compounds will have compelled the perfumer to study this complex assiduously; so much so that the following suggestions are not intended for him, but primarily for the student who will find them an excellent basis for experiment. The formulae appended contain only the vital constituents of each type and may be elaborated, and perhaps improved, by the use of other aromatics selected from the odour classification. It should be borne in mind, however, that cost is the prime factor in this complex, and will depend largely upon the percentage of genuine otto employed.

Rhodinol is the outstanding rose alcohol in this gamut owing to its powerfully sweet odour, but the finest qualities are expensive due to the high cost of Bourbon geranium oil. Geraniol from palmarosa oil is the next raw material of importance, and although it possesses an intense sweetness it lacks the body of the former. Citronellol and cinnamic alcohol, too, are excellent constituents, together with the esters of the four alcohols, but the acetate and formate are usually preferred on account of their clean rose character. Other indispensable aromatics are phenyl ethyl alcohol and its esters, but it is doubtful if the latter offer any great advantages over and above the straight alcohol. There are two other substances that are common to many rose compounds: trichlor phenyl methyl carbinyl acetate and ionone, and there the list ends; for any further additions modify the note to that of a particular type of flower. The persistent rosy character of the crystalline synthetic favours its generous application, but ten per cent of ionone is about the normal limit, because if larger quantities are blended with rhodinol the note of the finished compound will have a distinct resemblance to the fragrance of red roses.

Since the honey-like sweetness of the Bulgarian oil is a unique feature of this complex special attention must be paid to its duplication, and while phenyl acetic acid is most useful for imparting this particular quality to such compounds, a less common synthetic, paramethyl quinoline, should not be overlooked. Moreover, the intensification of this characteristic is facilitated by the employment of small percentages of phenyl acetic aldehyde and of undecylenic aldehyde, while an overall richness may be imparted by traces of eugenol or methyl eugenol. However, the recent discovery and synthesis of damascenone, a ketone found in Bulgarian oil by Firmenich, will perhaps, when available, prove to be the essential constituent of compounds by imparting character impact and closer duplication of the natural oil.

FLOWER PERFUMES: ROSE 195

The following formula illustrates these suggestions, and for the best results it should be diluted with the maximum quantity of genuine otto allowed by costs.

Synthetic Rose Otto, no. 1056

200	Citronellol
100	Phenyl ethyl alcohol
200	Geraniol palmarosa
200	Rhodinol
50	Guaiac wood
5	Eugenol
70	Ionone alpha
50	Cinnamic alcohol
40	Phenyl acetic acid
3	Phenyl acetic aldehyde
80	Trichlor phenyl methyl carbinyl acetate
2	Undecylenic aldehyde
1000	

The red rose complex is dominated by equal quantities of rhodinol and/or its esters and of ionone, to which both phenyl ethyl alcohol and cinnamic alcohol add the required modification. But the richness of jasmin will be apparent on smelling this flower, and the addition of either the absolute or a good compound will provide this tonality to complete the blend, as follows:

Red Roses, no. 1057

150	Phenyl ethyl alcohol
200	Rhodinol
300	Ionone
150	Cinnamic alcohol
100	Rhodinyl acetate
50	Trichlor phenyl methyl carbinyl acetate
50	Jasmin compound, colourless
1000	

The duplication of the fragrance of white roses is a more complicated problem, because although the overall perfume is characteristically rosy there is a deep and full background odour requiring further study which, however, may be matched by the judicious use of patchouli, santal and vanillin. The following example will suggest the key:

White Roses, no. 1058

100	Benzyl acetate
100	Phenyl ethyl alcohol
20	Orris concrete
100	Ionone alpha
50	Ylang
30	Rose absolute
70	Rhodinyl formate
5	Civet absolute
20	Hydroxy citronellal
5	Musk ketone
30	Patchouli
10	Phenyl acetic aldehyde
50	Santal
10	Vanillin
150	Jasmin compound, colourless
250	Synthetic rose otto
1000	

A finished perfume may be prepared from the above as follows:

Rose Perfume, no. 1059

20	Synthetic rose otto
70	Red roses compound
50	White rose compound
2	Rose otto, Bulgarian
5	Rose absolute
2	Jasmin absolute
1	Civet absolute
50	Musk tincture, 3 per cent
800	Alcohol
1000	

The creation of special types is best accomplished by adding the necessary modifiers to the basic rose oil, and in almost all cases *traces* will be found sufficient. The following list will indicate the lines on which to proceed:

R. Banksian	Alpha ionone or orris oil.
Canina	Basil oil or ethyl decine carbonate.
Desprez	Ethyl cinnamate or rhodinyl butyrate.
Eglantine	Benzyl acetate and hydroxy-citronellal or phenylethyl acetate or iso-butyl phenylacetate:

Marechal Niel	Guaiac-wood oil and undecalactone or iso-butyl acetate.
Maschata	Iso-eugenol.
Muscosa	Oakmoss resin (colourless).
May	Cinnamic aldehyde.
Socrates	Undecalactone.
Unique jaune	Phenylacetic aldehyde or cinnamic alcohol.

Sweet pea

History

Classical literature contains no references of importance to this very charming garden plant. Theophrastus mentions a few field peas, and there is one reference to tine-tare, *Lathyrus tuberosus*, but in this case it occurs when he is describing various roots. The origin of the plant as we know it today, with its delightful perfume, appears to be somewhat obscure. It is generally believed to be a native of the island of Sicily where it is occasionally found growing wild. Its introduction to English horticulture appears to be in the early part of the eighteenth century. In the language of flowers, sweet peas are today emblematic of 'delicate pleasures'.

Varieties

The genus *Lathyrus* comprises a number of hardy annual and perennial plants much appreciated for decorative purposes in our gardens. They belong to the N.O. Leguminosae and the sub-order Papilionaceae and are closely allied to the Vetches. Botanically the sweet pea is *L. odoratus*, but in common parlance other varieties often bear this description. Some of the more interesting of these are as follows:

L. latifolius. The common everlasting pea found throughout Europe. Flowers rose-coloured and some white and striped, 4 to 5 feet high, blossoms July and August.

L. grandiflorus. Large flowered or two flowered everlasting pea, common to southern Europe. Purplish-rose flowers appearing at their best in June.

L. rotundiflorus. Round leafed everlasting pea. Persia and Asia Minor. Flowers bright crimson-red. June to August.

L. tuberosus. Tine-tare, Dutch mice, Fyfield pea, tuber pea.

Common throughout Europe. Rather short and rambling. Bright rose flowers. July and August.

Odour

The sweet pea bears elegant and many coloured flowers of delicate and sweet fragrance. Their odour recalls that of orange-blossom and hyacinth with a suggestion of rose.

Natural perfume

This is not an article of commerce, although there is every reason to believe that enfleurage would capture the flower perfume faithfully. Since the duplication of the odour is fairly easily attained by the skilful blending of synthetics, it is doubtful whether the production of the natural perfume would meet with any great demand. The composition of this is not known, but the odour would at once suggest the presence of methyl anthranilate.

Odour classification

Top notes	Middle notes	Basic notes
1. Benzyl acetate	15. Guaiac wood	65. Cinnamic alcohol
2. Linalol	Heliotropin	77. Methyl naphthyl ketone
Mandarin	Styrax oil	
Phenyl ethyl acetate	21. Anisic aldehyde	80. Hydroxy citronellal
	Ionone alpha	
3. Terpineol	Methyl anthranilate	91. Undecalactone
4. Phenyl ethyl alcohol		100. Amyl cinnamic aldehyde
5. Ethyl phenylacetate	22. Geranyl formate	
6. Bergamot	23. Broom absolute	Benzoin resin
Iso-butyl phenylacetate	24. Jonquille absolute	Decyl aldehyde
	Rose otto	Methyl nonyl acetic aldehyde
7. Geraniol, Java	Ylang	
Rue	31. Orange flower absolute	Musk ambrette
8. Nerol		Phenyl acetic aldehyde
11. Narcissus absolute	43. Jasmin absolute	
Sweet orange	Tuberose absolute	Styrax resin
12. Methyl heptine carbonate	50. Neroli	Tolu balsam
Petitgrain, French		Vanillin

Compounding notes

The delicate fragrance of sweet pea is always appreciated in the home, but strangely enough the perfume has never appealed to the imagination of the public. Nevertheless, a study of its synthesis is desirable herein, and since the odour recalls that of a blend of orange blossom with hyacinth no great difficulty should be encountered in its duplication. In fact, a pleasing result may be obtained by direct admixture of these two compounds, flowered with neroli and intensified with decyl aldehyde, but this would leave something to be desired as an accurate duplication of the perfume, and in the circumstances it might be better to build up the complex from its unblended constituents. The greenish top note may be secured by blending linalol with petitgrain and shading with benzyl acetate, iso-butyl phenylacetate and methyl heptine carbonate. The essentials of the middle note are methyl anthranilate and neroli, together with broom absolute to which may be added ylang to taste. The indispensable constituents of the basic note are methyl naphthyl ketone, hydroxy citronellal and phenyl acetic aldehyde, which may be further intensified and modified with undecalactone, amyl cinnamic aldehyde, decyl and methyl nonyl acetaldehyde. The odour classification offers endless variations, but the above suggestions incorporating in the following formula will make an excellent starting point for experiment.

Sweet Pea, no. 1060

70	Benzyl acetate
200	Linalol
40	Iso-butyl phenylacetate
10	Methyl heptine carbonate
100	Petitgrain, French
80	Methyl anthranilate
20	Broom absolute
100	Ylang
50	Neroli
50	Methyl naphthyl ketone
150	Hydroxy citronellal
1	Undecalactone, 10 per cent
20	Amyl cinnamic aldehyde
5	Decyl aldehyde, 10 per cent
4	Methyl nonyl acetaldehyde, 10 per cent
100	Phenyl acetic aldehyde
1000	

A finished perfume may be prepared as follows:

Sweet Pea, no. 1061

140	Compound, no. 1060
1	Rose otto
5	Jasmin absolute
3	Orange flower absolute
1	Civet absolute
20	Ambergris tincture, 3 per cent
30	Musk tincture, 3 per cent
800	Alcohol
1000	

Trèfle

History

Trefoil appears to have been known from time immemorial, and Pliny[22] describes three varieties, the leaves of which were much used for making chaplets. In another part[23] of his voluminous works he mentions four remedies derived from the plant, the most important being presumably that it will cure the stings of serpents and scorpions.

Clover abounds in Ireland, and thus the strong belief has come about that this plant is the traditional shamrock. Some botanists, however, are of the opinion that the emblem was obtained from the wood sorrel. According to the legend, the Pope Celestine sent St Patrick to preach the gospel to the Irish. He landed at Wicklow in A.D. 433 and explained the doctrine of the Trinity by alluding to the triple leaflets of a plant he took from the ground. (It may have been *T. minor*.)

The plant

The common red clover *Trifolium pratense*, N.O. Leguminosae, is grown extensively in our pastures as a food for cattle. There are several varieties, by some botanists considered as distinct species, and many of them are mentioned by Theophrastus and other ancient writers. The most popular garden variety is *T. incarnatum*, the carnation or scarlet clover, and the odour exhaled by its charming flowers has led to the creation of many perfumes bearing

[22] Book XXI, chap. 30. [23] Chap. 88.

the name. Another variety possessing a beautiful honey-like fragrance is *T. odoratum*, a native of Italy. Other varieties are:

> *T. medium*, the zig-zag clover.
> *T. hybridum*, the alsike clover.
> *T. repens*, the white or Dutch clover.

T. pennsylvanicum, a native of the United States, resembling the latter, and extensively grown in this country, large quantities of seeds being imported annually from America. During the daytime the follicles of clover are fully expanded, but as night approaches they fold back, umbrella fashion, on to the stalk.

Natural perfume

The odour of clover is not extracted on a commercial scale, and the trèfle perfumes marketed by so many firms contain no real clover perfume. The floral note is in all cases obtained with other flower absolutes. The blossoms all exhale a honey-like fragrance which in practical perfumery is based upon one or other of the salicylates.

Chemistry

The chemistry of this perfume has so far received very little attention. H. Rogerson distilled the flowers of *T. incarnatum* and obtained 0·029 per cent of a powerfully odorous pale yellow volatile oil. He was able to identify furfural as a constituent. Power and Salway distilled *T. pratense* and obtained a yield of 0·028 per cent. The same aldehyde was identified in this oil.

Compounding notes

The Trèfle complex is one of the simplest in synthetic perfumery because the flowers are so obviously dominated by the odour of salicylates, toned with benzoates, that the problem resolves itself into the choice of the bases of these two esters and their ultimate blending to taste. Since the salicylates may be used up to as much as 60 per cent of the whole it is a moot point as to whether amyl or iso-butyl salicylate is the better, and moreover, as the benzoates are powerfully odorous substances the choice lies between ethyl and iso-butyl benzoate. Whichever of the above are decided upon the ratio of the first to the second will be about 9 to 1. Sweetening agents may be chosen from alcohols such as linalol, terpineol, geraniol and nerol. They are then flowered with ylang and rose and the whole intensified with small quantities of such basic elements as coumarin and musk.

Odour classification

Top notes	Middle notes	Basic notes
1. Benzyl acetate	15. Heliotropin	79. Civet absolute
2. Ethyl benzoate	20. Clary sage	80. Hydroxy citronellal
Linalol	21. Anisic aldehyde	87. Octyl aldehyde
3. Terpineol	22. Clove	91. Undecalactone
4. Citronellol	24. Ylang	100. Benzoin resin
Lavender	27. Dimethyl hydroquinone	Benzophenone
Nonyl aldehyde	43. Jasmin absolute	Coumarin
Phenyl ethyl alcohol	Rose absolute	Musk ketone
6. Bergamot	50. Neroli	Oakmoss
7. Geraniol, Java		Patchouli
8. Amyl salicylate		Phenyl acetic acid
Iso-butyl benzoate		Santal
Iso-butyl salicylate		Tolu balsam
Nerol		Vanillin
10. Linalyl acetate		Vetivert

The application of these constituents is illustrated in the following formula:

Trèfle, no. 1062

70	Ethyl benzoate
50	Linalol
50	Terpineol
600	Amyl salicylate
150	Nerol
4	Clove
50	Ylang
5	Rose absolute
1	Civet absolute
10	Coumarin
10	Musk ketone
1000	

A finished perfume may be prepared as follows:

Trèfle, no. 1063

140	Compound, no. 1062
1	Clary sage
4	Jasmin absolute
4	Rose absolute
1	Oakmoss
50	Musk tincture, 3 per cent
800	Alcohol
1000	

Tuberose

History

Polianthes tuberosa, L., belongs to the N.O. Amaryllidaceae, and is believed to be a native of either Mexico or the East Indies. The origin of the plant's introduction into Europe is not very clear. As far as is known, Peiresc was the first cultivator of the flower in Provence. It is said that he sent Father Minuti, one of the Minim brothers, to Persia especially for it, and that the first bulb was planted in his garden at Beaugencier about 1652. Another version is that Trovar, a Spanish physician, imported the first bulbs from Ceylon in 1594, and from Spain cultivation spread to the south of France. The tuberose is at present largely cultivated in Italy for its bulbs, which form an article of trade and are exported to this and other countries. It was first cultivated in Britain about the year 1630, and is sometimes called *Hyacinthus tuberosus* and *H. Indicus*. It derives its generic name from *polis* a city and *anthos* a flower, literally, flower of the city, and should not be confounded with polyanthus of the primrose family or with *Narcissus polyanthus.*

The tuberose has for years been regarded as the symbol of voluptuousness, and the reasons for this may be traced to the beliefs of some of the older writers who generally considered the perfume to be slightly intoxicating. For instance, one writer recommends good girls not to breathe the odour of the tuberose on a fine evening, because its subtle perfume throws one into a voluptuous intoxication from which one does not easily become liberated.

The Plant

The plant has a tuberous rootstock, from which grows an erect stem, 3 or 4 feet high, with a few slender long leaves and bearing from 14 to 16 pairs of flowers. In the south of France (the principal centre of cultivation being Pegomas) the bulbs are planted in April, about 1 foot apart, and between each row is left sufficient space for the collector to pass along and pick off each flower as it opens (fully opened flowers do not yield continuously to enfleurage but wither and spoil the pomade). This occurs about mid-day, and the flowering season lasts from the end of July to the early part of October. In November the roots are removed from the ground and stored in dry sand to preserve them.

There are several varieties of tuberoses. 1. The large flowered kind, which is the one generally cultivated for the extraction of its perfume. 2. The small flowered variety. 3. The streaked-leaf form. 4. The double-flowered variety grown from seeds by Lecours about 1760 at Leyden. 5. 'The Pearl', grown by Henderson in the States in 1865. 6. *Polianthes Blissii*, a hybrid grown by Blin in 1905 by crossing *P. tuberosa* with *Bravoa geminiflora*. Other varieties are known and include Albino and Excelsior.

Odour

The flowers of tuberose emit a most delightful fragrance, which in itself is a wonderful bouquet, and has been compared with the perfume of a well-stocked flower garden at evening close. In the East these white tubula flowers emit a most powerful odour which increases after sunset, and on this account is called by the Malays 'Mistress of the Night'.

Natural perfume

This is extracted generally by enfleurage, as the flower continues for forty-eight hours to produce odoriferous bodies while in contact with the fat. This pomade is extracted with alcohol to produce the 'Absolute from Pomade', 150 kilos of flowers producing about 1 kilo. The flowers are also extracted direct by means of volatile solvents, and the *absolute* obtained by this means is one of the most expensive natural flower products; 1200 kilos of flowers produce 1 kilo of concrete which in turn yields only about 200 grams of absolute. The yield by these two processes differs largely, and according to experiments conducted by Hesse the former method produces 13·32 times more essential oil than the latter. Niviere is of the opinion that the larger yield to enfleurage is subject to variation according to the percentage of lard employed. The present astronomic price of absolute has severely reduced its production.

Chemistry

The volatile oil has been analysed, but no quantitative figures have been published. Among those constituents identified by Hesse and Verley are the following:

> Benzyl alcohol
> Benzyl benzoate
> Methyl anthranilate
> Methyl benzoate
> Methyl salicylate
> A ketone, called tuberone

The methyl anthranilate is believed to average about 2 per cent and the ketone 10 per cent. Elze made a further examination of

Odour classification

Top notes	Middle notes	Basic notes
1. Benzyl acetate Methyl benzoate	15. Cinnamyl formate Eugenol	65. Cinnamic alcohol 80. Hydroxy citronellal
2. Benzyl formate Linalol Methyl salicylate	21. Methyl anthranilate 22. Geranyl formate 24. Ylang	100. Amyl cinnamic aldehyde Benzoin resin
6. Bergamot Iso-butyl phenylacetate	43. Jasmin absolute Tuberose absolute 50. Laurinic aldehyde	Coumarin Decyl aldehyde Gamma nonyl lactone
7. Geraniol, Java	Neroli	Iso-eugenol
8. Benzyl salicylate Iso-butyl benzoate Nerol	59. Celery seed	Labdanum resin Musk ambrette Peru balsam
12. Methyl heptine carbonate Phenyl ethyl cinnamate		Tolu balsam Vanillin
13. Paracresyl phenylacetate		

tuberose oil, steam distilled for the absolute, and in addition to the above-mentioned constituents found:

>Eugenol
>Geraniol
>Nerol
>Farnesol

The alcohols were combined in part with acetic and propionic acids.

Compounding notes

When a large bunch of the wax-like blossoms of tuberose is placed in a warm, closed room the perfume soon becomes heavy and almost overpowering; so much so that its aphrodisiac qualities are apparent and on this account it is disliked by many. Nevertheless, this fragrance has great aesthetic beauty and at one time its natural essence found wide use in perfumery. But today the astronomical cost of the absolute has compelled manufacturers to discontinue its use and to adopt in its place synthetic oils of fine quality. The duplication of the perfume is by no means easy, but those who smell the flowers analytically will notice that they have a fragrance of orange blossom type, combined with an overall suggestion of coconut, and once this has been observed with certainty it follows that the complex must be based upon methyl anthranilate, hydroxy citronellal and gamma nonyl lactone. But these aromatics alone will never satisfy the connoisseur and further study of the flower is necessary to elucidate the problem. However, among the identified constituents are methyl benzoate and methyl salicylate, both of which when blended with benzyl acetate and shaded with methyl heptine carbonate will modify the top note in the direction of the true fragrance. Subsequent flowering with natural tuberose and ylang will improve the blend but not complete it accurately, the final secret of which is celery. Thus, these are really all the vital constituents of a first class compound, to which some perfumers prefer to add tolu balsam and a trace of vanillin for perfection.

These suggestions are illustrated in the following formula:

Tuberose, no. 1064

100	Benzyl acetate
40	Methyl benzoate

70	Linalol
40	Methyl salicylate
100	Benzyl salicylate
10	Methyl heptine carbonate
200	Methyl anthranilate
40	Ylang
20	Tuberose absolute, from pomade
9	Celery
300	Hydroxy citronellal
50	Nonyl lactone
20	Tolu balsam
1	Vanillin
1000	

A finished perfume may be prepared thus:

Tuberose, no. 1065

100	Tuberose compound
40	Gardenia compound
6	Jasmin absolute
3	Orange flower absolute
1	Musk ambrette
40	Ambergris tincture, 3 per cent
10	Musk tincture, 3 per cent
800	Alcohol
1000	

Violet

History

Viola odorata is a perennial herb of the N.O. Violaceae, native of Britain, and found growing in woods and on the banks of our waysides. It appears to have been a great favourite even as far back as the pre-Christian era, for we find references to it in the works of Theophrastus (400 B.C.) and Ovid (43 B.C.). The latter has given us a story in Book I of his 'Metamorphoses', from which we gather that the violet took its name from Io, the daughter of Inachus. She was loved and ravished by Jupiter and changed by him into a heifer to avoid detection by Juno. Flowers (presumably violets) were caused to grow in the pastures around her in order that her grass might be distinguished from that of other animals. She was driven over the world by a gadfly sent by Juno, and at last came to the banks of the Nile, where she regained her human form. Her son was worshipped in Egypt as Isis. The violet is also mentioned by Horace, Pliny, and Juvenal the Roman satirist. Pliny called the white violet the 'first messenger of spring', and he so esteemed its perfume that he placed it immediately after roses and lilies. At that time the purple violet was called 'Ion' after the Greek, and from it ionthine (violet-coloured) cloth took its name. Pliny describes seventeen remedies derived from this flower. Mahomet, the prophet, appears to have gone further than this, for he compared the flower to the superiority of El Islam above other religions, while in the Koran he suggests that his superiority over other men is like that of the violet perfume over all others. About the year 1600 Herrick and Rapin enlarged on the possible origin of the flower, but their theories do not appear as probable as the one given above. The violet is the emblem of the Bonapartes, the flower having been adopted by the followers of the first Napoleon when in exile. He was styled Père la Violette, and a small bunch of violets hung up in the house or worn by a Frenchman denoted the adherence of the wearer to his fallen chieftain's cause. Within comparatively recent years the violet has been valued on account of its supposed medicinal qualities. The roots were stated to have emetic properties, and on that account to have been used as an adulterant of ipecacuanha (?). The leaves were used in the form of an infusion, which was applied both internally and externally as a supposed cure for cancer, while even today a syrup made from the petals (?) is an ordinary galenical obtainable from any pharmacist.

Sweet violets are nowadays highly prized for their delightful fragrance, both in Europe and the East. In the south of France they are cultivated for the extraction of their odoriferous constituents, and the blossoms are collected from the beginning of the year to March and April, while in the Middle East the flowers bloom with the hyacinth and narcissus in January, and are used by the native women to adorn their dark tresses.

Varieties

According to Sawer there are nine distinct varieties of *Viola odorata* L., both single and double, which are finely perfumed, and these include the following:

Devon, dark blue and very fragrant.
Neapolitan, lavender blue, large double flowers.

The Czar, large deep violet flowers, delightfully perfumed.

Victoria, single deep blue flowers on long stems.

Victoria Regina, larger flowers of a rich violet-blue, and most odoriferous.

In the south of France (Grasse and Hyères districts) Parma and Victoria violets are cultivated and sold as cut flowers, being sent to the capitals of Europe. A variety of Victoria violets known as *Luxonne* is the only kind of this species cultivated for perfumery purposes.

Odour

It is doubtful if the perfume of any other flower is as popular as that of the violet, and the odour of cassie blossom is the only one which bears any marked resemblance to it. Orris and costus roots when dried develop an aroma which is similar.

An odour of violet type has been noticed also in the following:

Flowers of *Ochrocarpus siamensis*, T. Anders, an Indian tree belonging to the N.O. Guttiferae and known locally as 'Tharapu'.

Flowers of *Pergularia odoratissima*, Sm., a tree belonging to the N.O. Asclepiadacea and found in Hawaii. It is there known as 'Chinese Violet' or 'Pakalang' and the odour is sometimes extracted by greases.

Flowers of *Securidaca longepedunculata*, a tree of the N.O. Polygalaceae, found growing in south-west Africa, where it is known as the 'Violet Tree'.

Natural Perfume

In the south of France violets are cultivated at Vence, Le Rouret, Opio, Roquefort, Valboune, Le Bar, Tourettes-sur-Loup, Magagnosc, and Chateauneuf. The total French crop of Victoria violets before the war was about 130,000 kg., and a larger crop of Parma violets was, however, available, but is now negligible. The quantity of wood violets, Czar Prince of Wales, and Wilson, has so diminished as to be also practically negligible. The propagation of the violet is by means of runners, stolons, seeds, and by separation of tufts. The suckers are separated from the parent plant towards the end of the winter or the early spring. They are planted out in rows about 2 feet apart, a space of 10 or 12 inches separating each plant. About mid-summer the lower parts are banked up to preserve them, and in October they are transplanted to the olive groves which affords the best protection from the cold in winter and the sun in summer. The ground is previously well manured with sodium nitrate 4, superphosphate 6, potassium chloride 3, and subsequently furrowed. The leaves appear in profusion by November, and the flowers bloom from early December onwards to sometimes as late as March or April; this, of course, being dependent on atmospheric conditions. The flowers and leaves are picked separately twice a week in the morning, and no time is wasted in transporting them to the factories for immediate treatment; a considerable part of their fragrance is lost if any delay occurs. Beds of Parma violets require four years to become commercially productive and Victoria violets two years. After six or seven years the plants become exhausted and are replaced. The flowers are extracted by maceration or by volatile solvents, and the absolutes from either of the resultant products are exceptionally expensive and only obtainable today when specially ordered. It is rather remarkable that the yield from both types of violets and their leaves is approximately the same. Some 4000 flowers weigh one kilo and the yield to petroleum ether is about 0·12 per cent of Parma concrete and 0·18 per cent of Victoria concrete, both varieties giving some 35 to 40 per cent of absolute. The concretes are green waxy materials having a comparatively weak odour, recalling both orris and carnation. Sabetay and Trabaud treated Parma violet concrete by steam distillation *in vacuo* and obtained a perfume content of 11·7 per cent. By the same means the absolute gave 32·7 per cent. Violet leaves yield to volatile solvent extraction 0·57 per cent of concrete giving 66 per cent of absolute. Owing to the astronomic price of Parma violet absolute it is very doubtful if any *absolutely pure* absolute is ever sold. Victoria violets are, by the way, much used in the manufacturing of confectionery.

Chemistry

Until a few years ago very little was known as to the constituents of violet flowers, and particularly of Parma violets, although violet leaves had been well studied from a chemical viewpoint. The absolute from the latter is characterised by a well-known intensely green odour, the main constituent being nonadienal which occurs to the extent of from 30 to 50 per cent. In addition the following are present: normal hexenic, heptenic and octenic alcohols nonadeine-2, 6-ol-1, hexyl and benzyl alcohols, and n-octene-2-ol-1 in the free state or in the form of propionic, oenanthic, octanoic, octenoic, palmitic and salicylic esters.

In view of the disagreement between the results of the chemical examination of violet flowers by H. von Soden, H. Walbaum and A. Rosenthal, the late Mr Fred Firmenich of the well-known firm in Geneva arranged for the preparation in Grasse of 700 grams of guaranteed pure absolute from Victoria violets. This oil was subjected by Dr Ruzicka and his co-workers to various tests for phthalic acid esters, violet leaf absolute, etc., all of which were negative, thus corroborating the purity of the product. Other alcohols in addition to nonadienal and benzyl alcohol were found in the violet flower absolute, amongst them n-hexyl alcohol, together with heptenol and octadienol, thus proving that the flower oil contains a number of alcohols closely identical to those in violet leaf. From olfactory tests it was quite evident that these various bodies were not the osmophores responsible for the odour that distinguishes violet flower oil. The true bearer of the characteristic odour of the flower perfume was not discovered until the Reagent T^{24} of H. Girard became available. By means of this reagent, Ruzicka was able to isolate a ketone which, even in very dilute solutions, had a strong resemblance to violet flowers. Empirical analysis established the formula $C_{13}H_{20}O$ for this new ketone.[25] Numerous tests demonstrated that this was a new isomer of the ionones, found only in violet flowers and absent from the leaves. The name *Parmone*[26] was given to this ketone, which is now synthesised in the Firmenich laboratories and is available in a compounded form as Parmantheme.

In their study of Parma violet absolute, Sabetay and Trabaud noticed its marked peppery odour and, according to their exhaustive tests, were able to prove the presence of 21 per cent of eugenol. They searched for this phenol in the absolute from violet leaves but only found trifling quantities recognisable by its odour. They did, however, find a high methoxy content which had hitherto escaped all investigations.

---

[24] Trimethyl carbohydrazidomethyl ammonium chloride.
[25] Irone is $C_{14}H_{22}O$.
[26] From the 700 grams of Victoria violet absolute, Dr Ruzicka was only able to isolate 0·25 gram of this ketone.

Violet roots when freshly bruised emit a fragrant smell, and since the presence of a volatile oil was suspected an examination was made by Goris and Veschniac. They were able to show that methyl salicylate was probably present.

Odour classification

Top notes	*Middle notes*	*Basic notes*
1. Benzyl acetate	15. Guaiac wood	73. Cassie absolute
2. Linalol	Heliotropin	79. Civet absolute
Phenyl ethyl acetate	16. Eugenol	80. Hydroxy citronellal
	18. Orris absolute	
3. Cuminic aldehyde	Violet leaf absolute	90. Orris resin
6. Bergamot	21. Anisic aldehyde	100. Benzoin resin
7. Methyl octine carbonate	Ionone alpha	Costus
	Ionone beta	Musk ketone
12. Methyl heptine carbonate	22. Benzyl iso-eugenol	Santal
	24. Rose otto	Vanillin
13. Orris concrete	Ylang	Vetivert
14. Methyl ionone	43. Jasmin absolute	
Mimosa absolute	Rose absolute	
Reseda absolute	50. Laurinic aldehyde	

Compounding Notes

The delicate and elusive fragrance of violets has always been popular, and despite the number of fancy bouquets that now flood the market this lovely perfume has never lost favour with those who delight in its unique qualities. There is only one criticism that can be levelled against it, and that is its peculiar property of fatiguing the olfactory system. It is well known that the Ionones dominate all violet compounds, and it is a matter of taste as to which of them are used to crown the complex. But it is essential to modify this odour by suitable additions of violet leaf absolute and/or methyl heptine carbonate, while the blend is finalised by the incorporation of jasmin absolute and/or its principal constituents and cassie absolute. Such a selection of aromatics will yield an excellent Parma violet, whereas the construction of Victoria violet requires a rebalancing of the above substances, together with a substantial increase in the percentage of ylang, plus other aromatics such as heliotropin and anisic aldehyde. Both types are illustrated in the following formulae.

Parma Violet, no. 1066

100	Benzyl acetate
100	Bergamot
10	Methyl heptine carbonate
20	Orris concrete
500	Methyl ionone
20	Violet leaf absolute
150	Ionone alpha
40	Benzyl iso-eugenol
20	Ylang
20	Jasmin absolute
20	Cassie absolute
1000	

Victoria Violet, no. 1067

100	Bergamot
10	Methyl heptine carbonate
10	Orris concrete
400	Methyl ionone
60	Heliotropin
10	Violet leaf absolute
40	Anisic aldehyde
100	Ionone beta
50	Benzyl iso-eugenol
60	Ylang
100	Jasmin compound
20	Cassie absolute
40	Hydroxy citronellal
1000	

A finished perfume may be prepared as follows:

Violet, no. 1068

100	Parma violet compound
40	Victoria violet compound
5	Cassie absolute
1	Rose otto
1	Civet absolute
1	Musk ketone
2	Santal
30	Ambergris tincture, 3 per cent
20	Musk tincture, 3 per cent
800	Alcohol
1000	

WALLFLOWER

The plant

Cheiranthus cheiri, popularly known as the giroflée, is a perennial plant of the N.O. Cruciferae and native of southern Europe. It was well known to the ancient Greeks, and is mentioned by Theophrastus (400 B.C.) in his *Enquiry into Plants.* The wallflower was introduced to this country nearly 400 years ago, and is now one of the commonest plants cultivated in our gardens. It is the flower with which romance-writers embellish their stories relating to ruins and desolate places. 'From the fact that wallflowers grew upon walls and were seen on casements and battlements of ancient castles and among the remains of abbeys, the minstrels and troubadours were accustomed to wear a bouquet of these flowers as the emblem of an affection which is proof against time and misfortune.' The plant seldom exceeds a foot in height, and in the natural state the flowers are single and of a yellowish colour. Cultivation has produced various beautifully coloured specimens, and many acres of them are grown to supply the London markets. The parent plant was called at one time the 'forty-day wallflower' because it often blossomed forty days after sowing. These should not be confused with the alpine wallflower, the fairy, and the rhoetian wallflower which belong to the genus *Erysimum.*

Chemistry

The natural perfume is not an article of commerce. The flowers were extracted by E. Kummert with volatile solvents, and after removal of natural waxes, etc., he obtained 0·06 per cent of an oil of disagreeable odour. This chemist established the presence of the following constituents:

Irone	Salicylic acid
Nerol	Anthranilic acid
Geraniol	Methyl anthranilate
Benzyl alcohol	Indole
Anisic aldehyde	Acetic acid
Linalol	and traces of phenols and lactones

Compounding Notes

The synthesis of the wallflower fragrance is based upon the liberal use of anisic aldehyde in combination with a rose alcohol, a salicylate and iso-eugenol. But the overall characteristic is not so easily duplicated, since there is a top note that recalls cuminic aldehyde and paracresyl methyl ether. These constituents are therefore indispensable but must be employed with great discretion otherwise they will ruin the complex. The best flowering agents are rose and jasmin, while undecylic aldehyde imparts a

Odour classification

Top notes	Middle notes	Basic notes
1. Benzyl acetate	15. Cinnamyl acetate	65. Cinnamic alcohol
2. Linalol	Heliotropin	73. Cassie absolute
3. Cuminic aldehyde	16. Eugenol	79. Civet absolute
Paracresyl methyl ether	21. Anisic aldehyde	80. Hydroxy citronellal
6. Bergamot	Indole	100. Amyl cinnamic aldehyde
7. Geraniol, Java	Ionone	
8. Amyl salicylate	Methyl anthranilate	Benzoin resin
Iso-butyl salicylate	22. Benzyl iso-eugenol	Decyl aldehyde
Nerol	24. Geranyl acetate	Iso-eugenol
12. Petitgrain, French	Jonquille absolute	Musk ketone
14. Methyl ionone	Rose otto	Peru balsam
	Ylang	Phenyl acetic aldehyde
	43. Jasmin absolute	Undecylic aldehyde
	Rose absolute	
	Tuberose absolute	

cachet that is inimitable. The following formula forms an excellent basis for experiment, with numerous opportunities of variation by choice from the odour classification.

Wallflower, no. 1069

100	Benzyl acetate
50	Linalol
5	Cuminic aldehyde
10	Paracresyl methyl ether
100	Amyl salicylate
100	Nerol
100	Methyl ionone
100	Heliotropin
200	Anisic aldehyde
5	Methyl anthranilate

Wallflower, no. 1069 (*continued*)

10	Rose otto
10	Jasmin absolute
100	Cinnamic alcohol
100	Iso-eugenol
10	Undecylic aldehyde, 10 per cent
1000	

A finished perfume may be made as follows:

Wallflower, no. 1070

140	Compound, no. 1069
5	Jonquille absolute
3	Rose absolute
1	Civet absolute
1	Musk ketone
50	Musk tincture, 3 per cent
800	Alcohol
1000	

CHAPTER SIX

Miscellaneous Fancy Perfumes

Non-alcoholic concentrates

Under this heading are included formulae which after maturing will be found useful for perfuming all kinds of toilet articles. In the event of their being required for perfuming brilliantines, etc., where turbidity would result on mixation, the essential oils should be replaced by their terpeneless equivalents and the natural absolutes by their artificial imitations.

Abronia, no. 1071

500	Verbena oil, French
50	Rhodinol
50	Ionone
20	Iso-eugenol
80	Heliotropin
40	Rose, synthetic
130	Bergamot oil
30	Musk ketone
50	Neroli synthetic
40	Benzoin R.
10	Cistus R.
1000	

Amber, Synthetic, no. 1072

250	Bergamot
30	Rhodinol
20	Vetivert
10	Patchouly

Amber, Synthetic, no. 1072 (*continued*)

100	Santalwood
70	Liquidambar Resin
10	Musk ketone
70	Coumarin
140	Vanillin
300	Ethyl phthalate
1000	

Ambre Royale Aux Fleurs, no. 1073

5	Diphenyl methane
5	Vetivert oil, English
5	Clary sage oil
10	Jasmin absolute
25	Violet, synthetic
25	Santalwood oil, English
25	Coumarin
50	Musk ketone
50	Heliotropin
50	Hydroxy-citronellal
50	Iso-butyl phenylacetate
50	Rose oil, Bulgarian
50	Amber, synthetic
50	Musk extract, 3 per cent
50	Benzoin R.
250	Benzyl benzoate
250	Ambergris extract, 3 per cent
1000	

Ambrosia, no. 1074

100	Coumarin
20	Dimethyl hydroquinone
20	Oakmoss, colourless
300	Iso-butyl salicylate
10	Methyl naphthyl ketone
200	Bergamot oil, terpeneless
200	Lavender oil, English
20	Ionone beta
5	Diphenyl oxide
25	Ylang-ylang oil
50	Jasmin, colourless
50	Iso-butyl benzoate
1000	

Benzoinette, no. 1075

5	Ethyl cinnamate
5	Anisic aldehyde
100	Vanillin
50	Coumarin
100	Tuberose, synthetic
200	Rose, synthetic
20	Vetivert oil
10	Patchouli oil
10	Oakmoss resin
100	Benzyl iso-eugenol
400	Benzoin R.
1000	

Boronia, no. 1076

200	Bergamot oil, terpeneless
50	Lemon oil
200	Rose, synthetic
10	Clary sage oil
90	Neroli, synthetic
50	Tuberose, synthetic
50	Heliotropin
50	Muguet, synthetic
200	Beta, Ionone
100	Boronia absolute
1000	

Bouquet Des Alpes, no. 1077

400	Lavender oil, French
100	Bergamot oil
100	Geraniol
5	Ambrette oil, 10 per cent
25	Ylang-ylang oil
5	Clary sage oil
25	Rosemary oil
50	Jasmin, synthetic
200	Benzoin R.
90	Ethyl phthalate
1000	

Bouvardia, no. 1078

80	Benzyl propionate
20	Benzyl iso-butyrate
100	Benzyl acetate
80	Cinnamic alcohol

Bouvardia, no. 1078 (*continued*)

130	Cedrat oil
20	Methyl anthranilate
200	Hydroxy citronellal
50	Alpha ionone
100	Linalol
70	Phenylethyl phenylacetate
20	Undecylic aldehyde, 10 per cent
30	Jasmin absolute
10	Rose absolute
10	Tuberose synthetic
50	Paracresyl phenylacetate
30	Heliotropin
1000	

Cananga, no. 1079

50	Jasmin, colourless
50	Metacresyl phenylacetate
20	Iso-eugenol
30	Phenylethyl formate
50	Neroli, synthetic
100	Jonquille, synthetic
675	Ylang, synthetic
25	Alcohol C_{10}
1000	

Coronilla, no. 1080

200	Hydroxy citronellal
150	Amyl phenylacetate
100	Coumarin
100	Dimethyl hydroquinone
100	Jasmin, colourless
50	Rose otto, synthetic
100	Rose, synthetic
200	Amber, synthetic
1000	

Corylopsis, no. 1081

300	Ylang-ylang oil, Manila
250	Benzyl acetate
15	Heliotropin
130	Rose, synthetic
50	Patchouli R.
5	Vetiverol
100	Hydroxy citronellal
150	Benzoin R.
1000	

Decumaria, no. 1082

500	Anisic aldehyde
25	Para-methyl acetophenone
100	Heliotropin
75	Vanillin
5	Aldehyde C_{16}, 10 per cent
95	Benzyl acetate
75	Petitgrain oil, terpeneless
25	Geranyl formate
100	Benzyl iso-eugenol
1000	

Dillenia, no. 1083

250	Terpineol
100	Rhodinol
250	Bergamot oil
50	Casie, synthetic
50	Orange blossom synthetic
30	Heliotropin
20	Musk ketone
250	Violet, synthetic
1000	

Erica, no. 1084

40	Diphenyl oxide
30	Oakmoss resin
200	Bergamot oil, terpeneless
100	Benzyl iso-eugenol
30	Dimethyl hydroquinone
50	Oleo-resin orris
150	Amber, synthetic
200	Geraniol
150	Citronellol
50	Coumarin
1000	

Fagonia, no. 1085

300	Terpineol
250	Linalol
150	Hydroxy citronellal
60	Ylang-ylang oil
50	Jasmin, colourless

Fagonia, no. 1085 (*continued*)

150	Geraniol
20	Vanillin
20	Civet absolute, 10 per cent
1000	

Glycine, no. 1086 (Wistaria)

180	Hawthorn, synthetic
50	Eugenol
100	Methyl ionone
100	Hydroxy citronellal
70	Ylang oil, Bourbon
80	Rose centifolia, synthetic
190	Jasmin, synthetic
100	Terpineol
40	Coumarin
60	Heliotropin
30	Musk ketone
1000	

Hancornia, no. 1087

180	Jasmin, synthetic
100	Tuberose, synthetic
50	Rose, synthetic
20	Cassie, synthetic
200	Benzyl propionate
350	Hydroxy-citronellal
25	Dimethyl hydroquinone
25	Clary sage oil
50	Styrax R.
1000	

Hugonia, no. 1088

100	Heliotropin
50	Vanillin
250	Ionone alpha
50	Orris oil, concrete
25	Casie synthetic
25	Jasmin synthetic
3	Aldehyde C_{12}
47	Santalwood oil
100	Tolu R.
50	Cistus R.
50	Musk ketone
250	Benzoin R.
1000	

Idealia, no. 1089

100	Bergamot oil
50	Sweet orange oil
80	Methyl ionone
20	Ylang oil, Manila
100	Carnation, synthetic
100	Chypre, compound
20	Cassie absolute
10	Neroli oil
170	Jasmin, synthetic
150	Rose, synthetic
50	Coumarin
40	Heliotropin
60	Vanillin
50	Amyl salicylate
1000	

Ismene, no. 1090

500	Muguet, compound
100	Ylang-ylang oil
50	Jasmin, synthetic
25	Tuberose, synthetic
23	Vanillin
100	Rhodinol
50	Olibanum R.
50	Opoponax R.
2	Undecalactone
100	Rose, synthetic
1000	

Jonesia, no. 1091

150	Bergamot oil, terpeneless
30	Lemon oil, terpeneless
20	Lime oil, terpeneless
100	Verbena oil, terpeneless
100	Neroli oil, North African
200	Bois de rose oil
200	Terpineol
30	Elemi R.
30	Myrrh R.
20	Benzyl formate
20	Geranyl formate
100	Benzoin R.
1000	

Kleinhovia, no. 1092

25	Mimosa absolute
50	Cassie absolute
25	Reseda absolute
50	Orange blossom synthetic
50	Jasmin synthetic
300	Violet, compound
200	Rose rouge, compound
50	Duodecyl alcohol
50	Concrete orris
200	Heliotropin
1000	

Laelia, no. 1093

200	Iso-butyl salicylate
300	Amyl salicylate
25	Oakmoss absolute
25	Labdanum R.
150	Coumarin
20	Dimethyl hydroquinone
180	Linalyl acetate
50	Lavender oil
5	Tarragon oil
5	Methyl heptine carbonate
40	Iso-butyl phenylacetate
1000	

Lime Blossom, no. 1094

400	Hydroxy citronellal
150	Terpineol
180	Phenylethyl alcohol
20	Ylang oil, Bourbon
10	Limes, oil, terpeneless
50	Methyl naphthyl ketone
30	Iso-Jasmone
70	Citronellyl acetate
5	Methyl octine carbonate
10	Iso-butyl phenylacetate
5	Aldehyde C_{12}
20	Jasmin absolute
10	Broom absolute
30	Heliotropin
10	Musk ambrette
1000	

Monimia, no. 1095

300	Grasse geranium oil
200	Ylang-ylang oil
100	Heliotropin
100	Vanillin
25	Iso-eugenol
50	Jasmin, synthetic
50	Patchouli oil
25	Oakmoss resin
30	Rose synthetic
20	Orange flower synthetic
100	Benzyl benzoate
1000	

Nemesia, no. 1096

125	Methyl ionone
100	Rose rouge compound
75	Cistus R.
100	Bergamot oil, terpeneless
100	Jasmin, colourless
50	Heliotropin
50	Coumarin
40	Musk ketone
260	Amber, synthetic
80	Linalol
20	Meta-cresol phenylacetate
1000	

Night Scented Stock, no. 1097

100	Mimosa, synthetic
100	Jasmin, synthetic
100	Methyl ionone
150	Citronellol
100	Geraniol
50	Bois de rose oil
50	Bergamot
100	Terpineol
20	Ylang Bourbon oil
20	Orris concrete
10	Jasmin absolute
10	Coumarin
70	Iso-eugenol
120	Heliotropin
1000	

Opoponax, no. 1098

400	Bergamot oil
70	Rhodinyl acetate
80	Santalwood oil
30	Citral
50	Musk ambrette
50	Heliotropin
30	Vanillin
70	Coumarin
30	Vetivert oil, English
20	Patchouli oil
70	Benzoyl acetone
100	Opoponax resin
1000	

Opoponax, no. 1099

300	Bergamot oil, terpeneless
50	Orris oil, concrete
50	Opoponax R.
20	Lemon oil, terpeneless
50	Jasmin, synthetic
80	Rose rouge, synthetic
10	Ginger oil
50	Myrrh R.
40	Galbanum R.
25	Encens R.
50	Violet, synthetic
25	Vetivert oil
50	Cistus R.
200	Amber, synthetic
1000	

Passiflora, no. 1100

370	Amber, synthetic
100	Coumarin
100	Heliotropin
200	Hydroxy-citronellal
5	Undecalactone
100	Dimethyl hydroquinone
100	Iso-butyl phenylacetate
25	Clary sage oil
1000	

Pavetta, no. 1101

40	Vetivert oil, English
60	Patchouli oil, English
100	Santalwood oil
200	Rose geranium oil
25	Musk extract, 3 per cent
25	Civet extract, 3 per cent
100	Benzyl iso-eugenol
100	Coumarin
50	Heliotropin
200	Bois de rose oil
100	Benzoin R.
1000	

Randia, no. 1102

400	Benzyl acetate
200	Hydroxy citronellal
50	Phenylacetic aldehyde
50	Ylang-ylang oil
100	Linalol
10	Aldehyde C_{11}
40	Alcohol C_{10}
50	Cinnamyl cinnamate
100	Cinnamic alcohol
1000	

Santolina, no. 1103

400	Lavender oil, French
100	Bergamot oil
75	Ylang-ylang oil
25	Clary sage oil
10	Oakmoss absolute
40	Ambrette R.
50	Vetivert R.
50	Cassie, synthetic
150	Rose, synthetic
100	Benzoin R.
1000	

Stephanotis, no. 1104

250	Heliotropin
250	Phenylethyl acetate
25	Paracresyl phenylacetate

Stephanotis, no. 1104 (*continued*)

40	Coumarin
60	Rhodinol
25	Musk ambrette
10	Civet extract, 3 per cent
30	Cassie synthetic
30	Jasmin synthetic
30	Jonquille absolute
50	Tuberose synthetic
200	Amber, synthetic
1000	

Syringa, no. 1105

250	Hydroxy-citronellal
100	Terpineol
50	phenylacetaldehyde, 10 per cent
100	Anisic aldehyde
200	Jasmin extra, compound
100	Orange blossom, compound
10	Acetophenone
90	Cinnamic alcohol
100	Heliotropin
1000	

Tinnea, no. 1106

50	Duodecyl alcohol
50	Ethyl decine carbonate
100	Ionone
200	Methyl ionone
100	Benzyl iso-eugenol
100	Santalwood oil
100	Ylang-ylang oil
50	Musk ketone
50	Vanillin
200	Rose, compound
1000	

Well-known recipes

The following formulae represent the approximate equivalent for many popular alcoholic handkerchief perfumes, recipes for which have been published.

Experiment will show the proportion of spirit required for their dilution.

A la Mode, no. 1107

15	Benzaldehyde
35	Nutmeg oil, expressed
120	Cassie
350	Jasmin
220	Orange blossom
220	Tuberose
40	Civet extract, 3 per cent
1000	

Bouquet des Fleurs, no. 1108

240	Bergamot oil
120	Lemon oil
120	Portugal oil
150	Violet
150	Tuberose
150	Rose
70	Benzoin R.
1000	

Bouquet d'Esterhazy, no. 1109

250	Neroli
250	Orange flower
250	Rose
50	Eugenol
50	Santalwood oil
40	Coumarin
10	Vanillin
10	Vetivert oil, English
40	Concrete orris
50	Amber, synthetic
1000	

Buckingham Flowers, no. 1100

200	Rose
200	Jasmin
200	Cassie
200	Orange flower
50	Concrete orris
50	Rose otto, synthetic
30	Neroli oil petale
30	Lavender oil, English
40	Amber, synthetic
1000	

Essence Bouquet, no. 1111

160	Petitgrain oil
160	Sweet orange oil
140	Rose
130	Jasmin
120	Lavender oil, French
80	Lemon oil
80	Neroli
25	Rose oil, Bulgarian
25	Thyme oil, white
20	Palmarosa oil
20	Clove oil
20	Cassia oil
20	Musk extract, 3 per cent
1000	

Eau de Berlin, no. 1112

40	Aniseed oil
480	Bergamot oil
20	Cardamon oil
15	Coriander oil
30	French geranium oil
30	Lemon oil
20	Melissa oil
75	Neroli oil
40	Rhodinol
30	Santalwood oil
20	Thyme oil, white
200	Ethyl phthalate
1000	

Frangipanni, no. 1113

60	Bergamot oil
200	Cassie
50	Civet extract, 3 per cent
80	Geranium oil, French
50	Musk extract, 3 per cent
70	Neroli oil
150	Orange blossom
40	Rose otto
200	Rose compound
10	Santalwood oil
60	Coumarin
30	Vanillin
1000	

Horse-Guard's Bouquet, no. 1114

700	Rose
150	Orange flower
60	Iso-eugenol
30	Concrete orris
20	Vanillin
40	Musk ambrette
1000	

Hovenia, no. 1115

340	Iso-eugenol
210	Lemon oil, terpeneless
225	Petitgrain oil, terpeneless
225	Rose
1000	

Japanese Bouquet, no. 1116

160	Cedarwood oil
160	Patchouli oil
160	Santalwood oil
160	Verbena oil, French
80	Vetivert oil
280	Rose
1000	

Jockey Club, no. 1117

160	Bergamot oil
200	Jasmin
300	Rose
140	Tuberose
160	Mace oil, concrete
40	Civet extract, 3 per cent
1000	

Kiss Me Quick, no. 1118

200	Cassie
50	Amber, synthetic
400	Jonquille
10	Paracresyl phenylacetate
20	Coumarin
20	Orris resin
40	Civet extract, 3 per cent

Kiss Me Quick, no. 1118 (*continued*)

200	Rose
40	Citronella oil
20	Lemon-grass oil
1000	

Leap-Year Bouquet, no. 1119

300	Jasmin
20	Patchouli oil
150	Santalwood oil
200	Tuberose
30	Verbena oil, French
150	Vetivert oil, Java
150	Rose
1000	

Bouquet a la Maréchale, no. 1120

100	Neroli oil, bigarade
200	Orange blossom
80	Coumarin
80	Vanillin
80	Orris oil concrete
100	Vetivert oil
200	Rose
40	Clove oil
40	Santalwood oil
40	Musk extract, 3 per cent
40	Amber, synthetic
1000	

Millefleur Bouquet, no. 1121

60	Cassie absolute
60	Methyl ionone
80	Jasmin extra
60	Neroli oil, bigarade
60	Patchouli oil
60	Vanillin
60	Violet
60	Vetivert oil
80	Lemon oil
120	Rose geranium oil
120	Lavender oil, French

80	Sweet orange oil
50	Musk extract, 3 per cent
50	Civet extract, 3 per cent
1000	

Musk, no. 1122

200	Musk ambrette
200	Musk ketone
50	Ionone beta
50	Vetiveryl acetate
100	Santalwood oil
400	Benzyl benzoate
1000	

Mousseline, no. 1123

150	Cassie
150	Jasmin
150	Rose de Provence
150	Tuberose
200	Bouquet à la maréchale
200	Santalwood oil, E.I.
1000	

Polyanthus, no.1124

300	Rose
150	Jasmin
75	Violet
200	Neroli
200	Lemon oil
75	Musk extract, 3 per cent
1000	

Rondeletia, no. 1125

10	Basil oil
120	Bergamot oil
250	Lavender oil
50	Clove oil
20	Rose oil, Bulgarian
300	Santalwood oil
40	Geranium oil, French
40	Cinnamon leaf oil
10	Ambrette oil, 10 per cent

Rondeletia, no. 1125 (continued)

20	Orris oil concrete
10	Jasmin absolute
10	Tuberose synthetic
70	Amber, synthetic
20	Vanillin
30	Musk ketone
1000	

Tulip, no. 1126

40	Cassie
260	Jasmin
135	Rose
260	Tuberose
260	Orris concrete
15	Benzaldehyde
30	Neroli
1000	

Yacht Club, no. 1127

200	Orange blossom
100	Jasmin extra
20	Cassie
200	Santalwood oil
50	Vanillin
100	Rose
330	Benzoin R.
1000	

Continental practice

The following formulae have been extracted from the works of the authors quoted, the same odour types being chosen so that comparisons may the more easily be made. Prior to 1912 the books published contained very few formulae utilising synthetics, and are therefore omitted. The formulae numbers have been appended by the author.

CONTINENTAL PRACTICE

1912. H. Mann, *Die Moderne Parfumerie.*

Chypre, no. 1128

10,000	Spirit
300	Infusion tolu balsam
300	Infusion Peru balsam
300	Infusion storax
100	Tincture musk
10	Turanol
150	Solution orris oil
100	Solution vetivert oil
5	Wintergreen oil
20	Aubepine
100	Bergamot
5	Citral
100	Lemon oil
5	Benzyl acetate
50	Geranium oil
20	Lavender oil
5	Eugenol
30	Cedarwood oil
11,600	

Lilac (Persian), no. 1129

7000	Tincture jasmin, 15 : 1000
5000	Tincture tuberose
5000	Tincture rose
200	Terpineol
30	Cananga oil
150	Infusion musk
30	Linalol
20	Vanillin
50	Muguet (Schimmel)
150	Infusion benzoin
20	Viodoran
17,650	

1912. R. Cerbelaud, *Formulaire de Parfumerie et de Pharmacie.*

Chypre, no. 1130

200	Infusion oakmoss
300	Infusion musk
25	Infusion civette
100	Infusion tuberose concrete, 2 per cent
100	Infusion jasmin concrete, 2 per cent
100	Infusion orange blossom, 2 per cent
200	Infusion orris, 20 per cent
1025	

Lilac (white), no. 1131

15	Terpineol
1	Vanillin
1	Musk ambrette
1	Ylang oil, Manila
2	Benzyl acetate
1	Linalol
1	Ionone
1	Cananga oil
25	Infusion ambrette seeds, 20 per cent
1000	Alcohol
1048	

1918. R. M. Gattefossé, *Agenda du Chemiste-Parfumeur.*

Chypre Imperial, no. 1132

10	Essence naturelle rose de Mai
5	Essence naturelle violet
5	Essence naturelle cassie
5	Essence naturelle orange flower
5	Essence naturelle tuberose
1	Essence naturelle orris concrete
5	Essence naturelle sauge sclarée
10	Essence naturelle mouse Evernia
12	Sassafras oil, terpeneless
5	Santalwood oil
10	Bergamot oil
2	Bois de rhodes oil
1	Clove oil
1	Canella oil, Chinese
0·1	Aniseed oil
10	Ambrette oil, concrete
10	Rose synthetic
10	Vanillin
5	Coumarin
5	Violet synthetic
117	

Lilac (white), no. 1133

600	Terpineol
150	Hydroxy-citronellal
50	Vanillin
20	Musk ketone
100	Benzoin resinodor
80	Jasmin absolute
1000	

1922. I. Lazennec, *Manuel de Parfumerie.*

Chypre, no. 1134

300	Tincture musk artificial, 0·7 per cent
2·5	Geraniol
100	Tincture ambrette seeds $\frac{1}{15}$
250	Tincture tonka beans $\frac{1}{10}$
150	Tincture vanilla $\frac{1}{10}$
150	Tincture orris
50	Alcohol
1002	

Lilac (white)

Duplicate of Cerbelaud.

1923. J. P. Durvelle, *Nouveau Formulaire des Parfums et des Cosmetiques.*

Chypre (for export)

Duplicate of Mann, but alcohol at 12 litres 250 cc.

Lilac (white), no. 1135

675	Alcohol
5	Terpineol
0·2	Ylang oil
1	Neroli synthetic
2	Jasmin essence
1	Rose essence
250	Infusion tuberose
200	Infusion jonquille
0·5	Ionone
0·2	Bitter almond oil
0·2	Clove oil
2·5	Infusion musk
1137	

1923. R. M. Gattefossé, *Formulaire de Savonnerie et de Parfumerie.*

Chypre, no. 1136

5	Clary sage
30	Oakmoss
20	Rose synthetic

Chypre, no. 1136 (*continued*)

20	Violet synthetic
5	Sassafras oil
5	Vetivert oil
3	Patchouli oil
3	Clove oil
10	Amber synthetic (in pieces)
20	Bergamot
20	Coumarin
141	

Lilac, no. 1137

200	Hydroxy-citronellal
250	Terpineol
100	Tuberique alcohol
200	Jasmin synthetic
750	

1924. L. Cuniasse, *Memorial du Parfumeur Chimiste.*

Chypre, no. 1138

5	Ionone
3	Civette artificial, 5 per cent
1	Musk ambrette
2	Geranium oil
50	Tincture ambrette seeds
10	Tincture labdanum, 50 per cent
100	Tincture orris
1	Vetivert oil
2	Santal oil
4	Patchouli oil
5	Sweet orange oil
4	Vanillin
10	Tincture Peru balsam
800	Alcohol
997	

Lilac, no. 1139

2	Neroli synthetic
2	Rhodinol
2	Linalyl acetate
5	Jasmin synthetic
25	Terpineol
5	Tincture benzoin

5	Tincture civette
0·2	Tincture bitter almond oil $\frac{1}{10}$
1	Vanillin
1000	Alcohol
1047	

1927. F. Winter, *Handbuch der Gesamten Parfumerie und Kosmetik.*

Chypre, no. 1140

15	Jasmin liquid, A
5	Rose liquid
6	Solution orris
6	Santal oil, E.I.
120	Bergamot
6	Patchouli oil
60	Musk ketone, $\frac{1}{5}$ in B.B.
5	Vetivert oil
200	Oakmoss tincture $\frac{1}{5}$
2	Coumarin
1·5	Vanillin
2	Heliotropin
25	Rose synthetic
10	Rose otto, Bulgarian
5	Pimento oil
10	Olibanum resinoid
4	Bitter orange oil
2	Ambrette seed oil
250	Musk tincture, 3 per cent
4000	Alcohol
4734	

Lilac, no. 1141

500	Terpineol
10	Ylang-ylang oil
15	Bromstyrole
300	Hydroxy-citronellal
20	Jasmin synthetic
40	Heliotropin
3	Bitter almond oil $\frac{1}{20}$
3	Peach lactone
891	

1928. A. Wagner, *Die Parfumerie Industrie.*

Chypre (French type), no. 1142

1320	Solution jasmin
1320	Solution rose
1720	Solution tuberose
1000	Solution cassie
133	Tincture musk
33	Civette synthetic, Heiko $\frac{1}{10}$
6·6	Calamus oil
26	Bergamot oil
20	Coumarin
6·6	Patchouli oil
6·6	Santal oil, E.I.
66	Castoreum tincture
13	Turanol
6·6	Iraldeine
13	Vanillin
666	Solution oakmoss
6356	

Lilac, no. 1143

300	Terpineol
300	Guaiac-wood oil
160	Iso-eugenol
160	Benzyl acetate
150	Cananga oil
150	Amyl salicylate
150	Phenylethyl iso-butyrate
65	Heliotrope synthetic
45	Ambra, A.
43	Phenylacetic aldehyde
18	Cyclosia
18	Lilac
15,000	Alcohol
1000	Water
17,559	

1929. H. Fouquet, *La Technique Moderne et les Formules de la Parfumerie.*

Chypre, no. 1144

25	Infusion oakmoss decolourised
35	Infusion tonka beans
45	Infusion orris
20	Infusion olibanum
10	Infusion civette

CONTINENTAL PRACTICE

15	Infusion ambrette seeds
25	Infusion vanilla
625	Extract 36, rose
75	Extract 36, cassie
75	Extract 36, jasmin
75	Extract 36, orange blossom
10	Essence rose, Bulgarian
15	Essence geranium, terpeneless
5	Essence jasmin
5	Essence santal
5	Essence neroli petales
2	Essence patchouli
2	Essence oakmoss
4	Essence vetivert
1	Essence clary sage
3	Methyl acetophenone
1	Octyl alcohol
2	Nonyl alcohol
2	Decyl alcohol
1	Octyl aldehyde
0·5	Nonyl aldehyde
0·5	Decyl aldehyde
1084	

Lilac, no. 1145

600	Extract 36, jasmin
150	Extract 36, tuberose
100	Extract 36, orange blossom
50	Extract 36, jonquille
5	Essence jasmin
5	Essence neroli petals
5	Essence Bulgarian
5	Essence ylang, extra
15	Linalol
15	Phenylethyl alcohol
50	Essence lilac synthetic, H.F.
20	Infusion vanilla
35	Infusion orris
10	Infusion styrax
15	Infusion civette
20	Solution civette artificial
1	Octyl alcohol
0·5	Nonyl alcohol
0·5	Nonyl aldehyde
9	Amyl cinnamic aldehyde
5	Tuberic alcohol
1116	

1930.　J. Broders, *Manuel du Parfumeur.*

Chypre, no. 1146

4	Infusion chypre
2	Infusion à la maréchale
2	Extract tuberose
1	Extract orange
1	Extract jasmin
1	Extract roses
11	

Lilac, no. 1147

1000	Infusion vanilla
2000	Infusion orris
4000	Extract orange
4000	Extract jasmin
2000	Extract tuberose
1000	Extract cassie
250	Infusion styrax
10	Lemon oil
10	Bergamot oil
1	Bitter almond oil
14,271	

1930.　A. M. Burger, *Leitfaden der Modernen Parfumerie.*

Chypre, no. 1148

Ingredient	Amount
Lemon oil	200
Bergamot oil	200
Sweet orange oil	200
Neroli oil	75
Amyl cinnamic aldehyde	150
Oakmoss	150
Rose oil, Bulgarian	50
Phenylethyl alcohol	100
Benzyl acetate	100
Cuminic aldehyde	s.

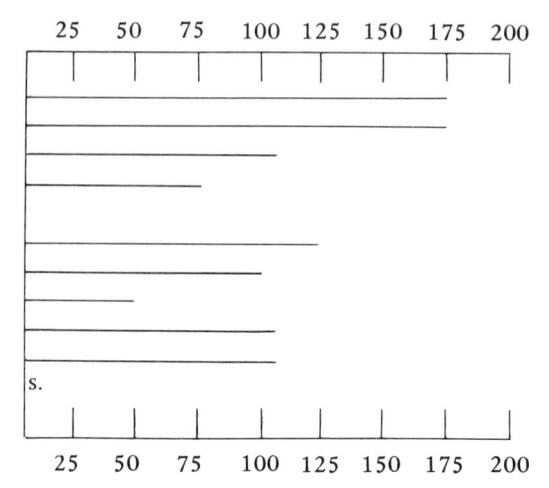

Lilac Base, no. 1149

Component	Amount
Heliotropin	25
Anisyl formate	25
Coumarin	10
Ylang, Manila	10
Terpineol	125
Hydroxy citronellal	125
Benzyl acetate	100
Amyl cinnamic aldehyde	40
Nerol	50
Linalol	50
Hydratropa aldehyde, 10 per cent	30
Phenylethyl alcohol	75
p-Methyl phenyl acetaldehyde	15
Methyl acetophenone	10
Indole	5

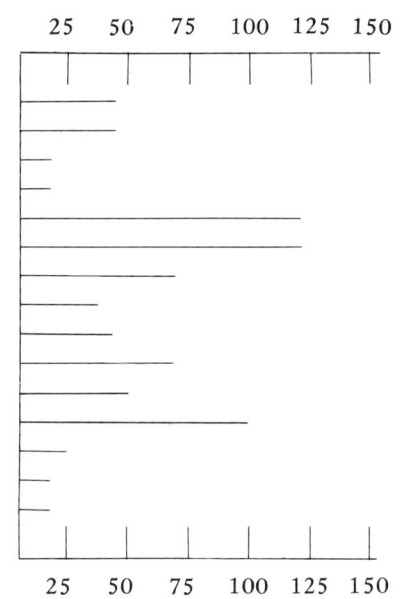

1931. O. Gerhardt, *Das Komponieren in der Parfumerie.*

Chypre Base, no. 1150 (Coty Type)

220	Santal, E. I.
227	Bergamot oil
50	Rose absolute
20	Fixonal
5	Coriander oil
50	Jasmin synthetic
40	Patchouli oil
7	Red thyme oil
110	Vetivert oil, Bourbon
55	Labdanum resinoid
110	Oakmoss absolute
70	Castoreum resinoid
20	Neroli synthetic
1	Iso-safrole
15	Musk ambrette
1000	

Lilac, no. 1151

448	Terpineol
133	Hydroxy citronellal
160	Heliotropin
50	Phenylethyl alcohol
82	Benzyl acetate
95	Anisaldehyde
6	Cananga
3	Coumarin
6	Alpha ionone
8	Jasmin aldehyde
6	Dimethyl hydroquinone
3	*p*-methoxy acetophenone
1000	

1931. F. Cola, *Le Livre du Parfumeur.*

Chypre, no. 1152

60	Santalol
90	Coumarin
30	Musk ketone
20	Musk ambrette
25	Ambreine absolute
25	Tarragon oil
5	Angelica root oil
30	Clary sage
60	Vetivert oil
30	Linalol
20	Patchouli oil
35	Iso-eugenol
50	Methyl ionone
60	Oakmoss absolute
225	Bergamot oil
20	Jasmin absolute
15	Rose absolute
2	Methyl salicylate
3	Lavender oil
15	Vanillin
35	Heliotropin
70	Ylang oil, Manila
25	Cinnamyl acetate
50	Benzoin resinoid
1000	

Lilac Bouquet, no 1153

165	Hydroxy citronellal
175	Cinnamic alcohol
155	Phenylethyl alcohol
10	Phenylacetic aldehyde
10	Anisic aldehyde, ex. anethol
45	Phenol glycol acetate
65	Heliotropin
20	Jasmin absolute
5	Rose absolute
5	Iso-eugenol
5	Para-cresol, 5 per cent solution
35	Benzyl acetate
40	Linalol
50	Phenylethyl acetate
10	Cuminic alcohol
10	Decyl alcohol, 10 per cent solution
10	Para-tolyl aldehyde, 10 per cent solution
5	Farnesol
120	Terpineol, extra rectified
5	Vanillin
15	Benzyl salicylate
45	Benzyl alcohol
1005	

CHAPTER SEVEN

Toilet Waters

Fragrant waters have been in use since the days of Theophrastus, and are believed to have hygienic qualities not possessed by ordinary handkerchief perfumes. It is difficult to explain their nomenclature, especially since they are generally made with weak alcohol, but, as the diluent usually consists of rose or orange flower *water*, it may be that this, to some extent at any rate, accounts for their designation.

Another distinction is, however, noticeable in nearly all modern formulae, and this is the absence of any flower *extract*, the principal constituents being either distillation or expression products. In the formulae which follow, it will be noticed that the *citrus* oils play an important part, even in lavender water, where the best effects are obtained by the liberal use of bergamot oil in conjunction with, of course, either English or French lavender oil.

Honey water

Honey water was probably the earliest member of this series, and is said to have been used by the ancient Greeks as a tonic for the hair. In later years it was prepared by distilling a mixture of honey, gum-arabic, and water, and was employed as a lotion for the face, which it made "white and fair". Nearly all modern formulae are based on the product originated by George Wilson, who manufactured it for King James II. His recipe for the preparation was approximately as follows:

No. 1154

100	Honey
100	Coriander fruit
9	Cloves
6	Nutmegs
6	Gum benzoin
6	Storax
6	Vanilla pods
10	Yellow lemon rind
243	

Bruise the cloves, nutmegs, coriander-seed, and benzoin, cut the vanilla pods in pieces, and put all into a glass alembic with a litre of French brandy, and after digesting forty-eight hours, distil. To one litre of the distillate add—

150	Damask rose water
150	Orange flower water
0·1	Musk
0·1	Ambergris
300	

Grind the musk and ambergris in a glass mortar, and afterwards put all together into a large matrass and let them circulate three days and three nights in a gentle heat; let them cool. Filter and keep the water in bottles well stoppered.

Honey water was popularised by Sir Erasmus Wilson, who prescribed it as a hair wash. The following formula will make a pleasantly perfumed product, such as is in demand today:

Aqua Mellis, no. 1155

5	Honey
8	Bergamot oil
1	Lavender oil, French
1	Clove oil
0·5	Nutmeg oil
1	Coriander oil
3·5	Santalwood oil
5	Benzoin R.
2	Musk extract, 3 per cent
100	Rose water triple
100	Orange flower water triple
800	Rectified spirit
1027	

This may be prepared as described under eau de Cologne, when an excellent product will result, but if distilling apparatus is not available, the whole should be macerated at least six months and afterwards filtered.

Hungary water

Hungary water is another *eau de toilette* of comparatively ancient origin, and was prepared mainly from rosemary. The fresh herb was taken and distilled with spirit, variations being sometimes made by the addition of lemon, lavender, or orris. There is very little call for this product today, but as it may be of interest the following formula is appended:

No. 1156

2	Rosemary oil
7	Verbena oil
1·5	Portugal oil
1	Limette oil
0·5	Peppermint oil
100	Triple rose water
100	Triple orange flower water
800	Alcohol, 90 per cent
1030	

Mature six months.

Eau de Cologne

Original type

Of all the toilet waters sold to the public, none are so popular as eau de Cologne, for it is known universally. There appears to be some doubt as to the actual origin of this perfume, and according to one version it was invented at Milan by Paul de Feminis, who manufactured it at Cologne in 1690. This gentleman is stated to have given the formula to his nephew, Jean Antoine Farina, who commenced to make it at Paris in 1806, and the manufacture of this particular eau de Cologne is supposed to be continued there today by a well-known firm. Another version, and probably the correct one, appeared in a well-known English periodical many years ago. According to this account J. M. Farina, who was the veritable inventor of what he called Kölnisches Wasser, or, as it is much more elegantly designated in its French synonym, eau de Cologne, was an Italian by birth, born at Santa Maria Maggiore, in the valley of Vigezza, district Domo d'Ossola, in the year 1685. He had emigrated to Cologne, however, and became a naturalised German, changing his first name to Johann at a somewhat early period. Certainly he was in business 'opposite the Julich's Place' in the year 1709, for his commercial books back to that date are still in the possession of the firm. Kölnisches Wasser is among the entries at that period, so that the perfume has been in existence certainly since that date. In 1726 the trade was flourishing, for in that year he sent for his brother John Baptist from Italy, who became his partner. The latter died in 1732, and John Maria, who was unmarried, found himself again alone. Then he sent for the son of John Baptist—who was also his own godson, and was luckily named John Maria—from Italy, and gave him a partnership. In 1766 the original old gentleman died, and left the concern exclusively to this John Maria the second. This one lived till 1792, after which his three sons—John Baptist, John Maria, and Charles Antony Hieronymus—reigned in his stead. The middle one of these, who was obviously intended to be the survivor of them all, perversely died first, and so for a moment the famous name was lost to the firm. But the other brothers both named their eldest sons John Maria, and these ultimately succeeded to the proprietorship of the business. The son of John Baptist died in 1833, but the other John Maria became head of the house. His son, who was also named John Maria, was actively associated with the senior Mr. Farina in the conduct of the business. The word Farina appears on several makes of eau de Cologne, and this is not surprising, since the name is a common one in Italy. At the present time there are two or three perfumers in Cologne who claim to be the original makers of this favourite toilet water. There is one characteristic about all the old-fashioned eau de Colognes, which is, that they represent a *type*, being, more or less, *citrus* bouquets blended with rosemary or lavender. They possess a refreshing and incomparable fragrance, which is typical of all the constituents.

The purity and source of the alcohol employed as a solvent for the oils is a factor which contributes materially to the odour of the finished perfume, and a perfectly neutral and highly rectified potato spirit is undoubtedly the most useful for this purpose. If this should not be available,[1] a treble-distilled molasses or grain alcohol will make a good substitute. Its method of preparation has been alluded to in the monograph on that subject in Volume I. The mere traces of oenanthic ether, which are present in these specially prepared raw materials appear to blend well with the oils and slightly modify their odour. The oils used should be selected from the few rather than the many "possibles", and may include neroli, petitgrain, lemon, orange, and bergamot, with the judicious additions of lavender and either rosemary or thyme.

Distillation also plays a most important role in the manufacture of "de luxe" products, but the oil of neroli should always be added afterwards. This process has a very subtle influence upon the fragrance of the constituents, and an entirely different and finer product results. The reasons for this peculiar change can only be conjectured, but that some molecular reconstruction of the essential oils takes place on distilling appears to be most probable. When distilling apparatus is not available the oils should be dissolved in the strong alcohol, and the water added little by little. The mixture is then placed aside in tanks when certain terpenes are precipitated. This process may be hastened by freezing, and immediate filtration yields a brilliantly clear liquid which will not cloud under any conditions. A reference to the odour classification which follows will suggest other basic notes, together with a long list of all aromatics that may be used in the creation of this complex. There are a few firms who still use rectified spirit for their toilet waters, but with the availability of better quality alcohol in Britain, a different complexion has been placed upon the manufacture of all perfumery. For the succeeding formulae the oils are merely added direct to the alcohol. Since the cost factor is of minor importance, maturing of the finished product for any period may be resorted to.

Products which closely resemble the original may be made as follows:

No. 1157

Place in a still—

- 8 Bergamot oil
- 6 Lemon oil
- 5 Sweet orange oil

[1] It is only obtained easily in Germany.

Odour classification

Top notes	Middle notes	Basic notes
1. Benzyl acetate 　 Ethyl acetate 　 Ethyl acetoacetate	15. Heliotropin 16. Serpolet 17. Melissa	77. Methyl naphthyl ketone 79. Civet absolute
2. Limes 　 Linalol 　 Mandarin 　 Octyl acetate 　 Rosewood	18. Marjoram 19. Verbena 20. Clary sage 21. Ionone alpha 　　 Methyl anthranilate	80. Hydroxy citronellal 90. Estragon 100. Ambergris 　　 Benzoin resin
3. Benzyl cinnamate 　 Coriander 　 Decyl formate	Myrrh resin 　　 Rosemary 22. Clove	Coumarin 　　 Decyl aldehyde 　　 Labdanum resin
4. Citronellol 　 Lavender 　 Phenyl ethyl alcohol	Geranyl formate 　　 Orange flower water absolute 24. Cinnamon bark	Musk ketone 　　 Tonka resin 　　 Vanillin
5. Terpinyl acetate 6. Bergamot 　 Citronellyl formate 　 Linalyl benzoate	Ethyl cinnamate 　　 Geranyl acetate 　　 Rose otto 　　 Ylang	
7. Geraniol, Java 　 Tansy 　 Thyme	26. Iso-butyl methyl anthranilate 30. Ambrette seed 　　 Cardamon	
8. Citronella, Ceylon 　 Citronellyl acetate 　 Geraniol 　　 Palmarosa 　 Lemon 　 Linalyl propionate 　 Nerol 　 Rhodinol	31. Orange flower absolute 35. Dimethyl anthranilate 40. Hyssop 41. Mace 43. Jasmin absolute 　　 Rose absolute 50. Neroli	
9. Neryl acetate 　 Peppermint	55. Nerolidol 60. Citral	
10. Lavandin 　　 Linalyl acetate		
11. Carrot seed 　　 Nutmeg 　　 Sweet orange		
12. Methyl cinnamate 　　 Petitgrain, French		
14. Basilic 　　 Lemongrass 　　 Methyl ionone		

No. 1157 (continued)

1	Lavender oil, French
10	Orris-root, crushed
500	Alcohol, 90 per cent
70	Water
600	

Macerate for twenty-four hours and then distil slowly.

Collect 500 c.c in the receiver.
To this add—

2·5	Neroli oil, bigarade
0·5	Rosemary oil
500	Alcohol, 90 per cent
5	Benzoin R.
508	

Mature one month.

No. 1158

5	Lemon oil
10	Bergamot oil
5	Portugal oil
1	Melissa oil
50	Rosemary herb
1000	Alcohol, 90 per cent
100	Water
1171	

Macerate for twelve hours in a still and then distil slowly.

Collect 1000 c.c., in which dissolve—

2	Neroli oil
3·5	Petitgrain oil, French
0·5	Thyme oil, rectified
4	Oleo-resin orris
10	

Mature one month.

No. 1159

5	Neroli oil
12	Bergamot oil

6	Lemon oil
1	Rosemary oil
0·5	Origanum oil
0·5	Lavender oil
50	Orange flower water triple
5	Essence ambergris, 3 per cent
950	Alcohol, 90 per cent
1030	

Dissolve the oils in the alcohol and add the essence.
Macerate seven days, and shake frequently.
Then add 10 c.c. of triple orange flower water each day for five days.
Filter bright, using talc if necessary.

Modern prototypes

Modern prototypes are legion, and a fault possessed by many of them is that they are overloaded with the essential oils already mentioned, and these are frequently supplemented by the addition of cinnamon, cloves, etc. While these alterations in formulae may be appreciated by some, they, nevertheless, so modify the odour that it is often difficult to identify the perfume as eau de Cologne. Such modifications are particularly noticeable in colognes for men, which today have a large sale and quite often yield a residual fragrance dominated by an oakmoss complex. Many delightful variations may be made without altering the type, and these are accomplished by the addition of rose, and by substituting—

Mandarin	for	Bitter orange
Petitgrain citronnier	for	Sweet orange
Coriander	for	Rosemary
Lime	for	Lemon
Clary sage	for	Lavender

With regard to the last example, it should be noted that clary sage oil is a magnificent and indispensable blender, and it is doubtful if any other individual essential oil exerts such an influence on the odour and tenacity of eau de Cologne. The general procedure should not be varied from that outlined above, but any of the Basic Notes mentioned in the odour classification may be added. The following formulae will indicate the lines upon which several attractive modern prototypes can be manufactured:

No. 1160

5	Lime oil, expressed
10	Bergamot oil
7	Portugal oil
1	Rose otto, Moroccan
1	Rosemary oil
1000	Alcohol, 90 per cent
100	Water
1124	

Distil 1000 c.c., in which dissolve—

2	Neroli oil
0·25	Clary sage oil
1	Benzyl iso-eugenol
3	

Mature one month.

No. 1161

5	Bergamot oil
6	Lemon oil
7	Petitgrain-citronnier oil
2	Rose geranium oil, French
1	Lavender oil
1000	Alcohol, 90 per cent
100	Water
1121	

Distil 1000 c.c., and add—

5	Petitgrain oil, terpeneless
1	Neroli oil, bigarade
2	Benzoin resin
8	

Mature one month.

No. 1162

1	Coriander oil
7	Lime oil, distilled
7	Mandarin oil
15	Bergamot oil
1	Thyme oil, white
0·5	Canadian snake-root oil
900	Alcohol, 90 per cent
200	Water
1131	

TOILET WATERS: EAU DE COLOGNE

Distil 1000 c.c., to which add—

0·5	Clary sage oil
1	Vetivert oil, English
0·5	Santalwood oil
8	Neroli, synthetic
10	

Mature three months.

No. 1163

0·5	Lime oil, terpeneless
5	Sweet orange oil
8·5	Bergamot oil
1	Verbena oil, French
800	Alcohol, 90 per cent
300	Water
1115	

Distil 1000 c.c., and add—

5	Rose geranium oil
5	Petitgrain oil, French, terpeneless
0·25	Ethyl cinnamate
5	Styrax R.
15	

Mature six months, and filter if necessary.

Amber Cologne, no. 1164

5	Lemon oil
10	Bergamot oil
5	Mandarin oil
5	Bois de rose oil
1	Rosemary oil
1000	Alcohol, 90 per cent
100	Water
1126	

Distil 1000 c.c., and add—

3	Coumarin
2	Vanillin
5	Benzoin resin

Amber Cologne, no. 1164 (*continued*)

0·5	Ethyl cinnamate
1	Clary sage oil
3·5	Neroli oil, bigarade
25	Essence ambergris, 3 per cent
40	

Mature one month.

Quickly matured colognes

Quickly matured colognes are best prepared by mixing the oils in bulk and allowing these to stand with frequent stirring for a month. Maturing is accelerated by once daily placing the container in warm water for an hour and then cooling. A matured concentrate thus saves tying up money by maturing for a long time in alcohol. By this method a reasonably good Cologne can be finished off after the oils and alcohol have been mixed a week. A type compound is as follows:

No. 1165

300	Bergamot oil
200	Lemon oil
150	Mandarin oil
50	Lavender oil, French
40	Rosemary oil
5	Clary sage oil
70	Neroli oil, bigarade
100	French petitgrain oil
10	Rose compound
20	Ethyl acetate
5	Musk ketone
50	Benzoin R.
1000	

Dissolve 5 per cent of this compound in 80 per cent alcohol, and after one week freeze and filter bright.

Flower modifications

Flower modifications enjoy a distinct popularity and may be classified as follows:

1. Those containing the traditional constituents together with the merest floral suggestion. The characteristic odour of eau de

Cologne only is at first apparent, but when the olfactory nerves have become accustomed to it, the flower, as it were, takes form, and the effect is both surprising and charming. Mimosa produces one of these most beautiful flower modifications, and will act as a good blender, especially if the natural absolute is used.

2. Those which do *not* contain any neroli or petitgrain, but have a very pronounced flowery odour. The perfume first recalls the flower, and afterwards eau de Cologne—in many cases there being no suggestion of the latter at all. This type is well represented by 'Trèfle Cologne', where the effect is produced by the use of iso-butyl salicylate, or amyl salicylate.

Mimosa Cologne, no. 1166

10	Bergamot
10	Lemon
10	Portugal
10	Petitgrain, terpeneless para
1	Coriander
1	Tarragon
3	Orange absolute
5	Mimosa absolute
100	Distilled water
850	Industrial methylated spirit
1000	

Mature one month, freeze and filter.

Trèfle Cologne, no. 1167

20	Bergamot oil
5	Lemon oil
1	Rosemary oil
1	Lavender oil, French
2	Ylang-ylang oil
1	Clary sage oil
1	Oakmoss resin
3	Musk extract, 3 per cent
1	Rose otto, synthetic
10	Iso-butyl salicylate
5	Coumarin
950	Alcohol, 80 per cent
1000	

Mature one month, freeze and filter.

Terpeneless oils

Terpeneless oils now play an important part in the manufacture of eau de Colognes, and their use enables the perfumer to employ alcohol of much weaker strength. Terpeneless and sesquiterpeneless oils should be chosen with care if really fine products are

Terpeneless oil	Quantity Dissolved by 1000 of Ethyl Alcohol at		
	80 per cent	70 per cent	60 per cent
Angelica	850	25	5
Bay	2500	500	150
Bergamot	1200	300	60
Caraway	1900	300	150
Cardamon	1300	450	80
Citronella	1450	20	3
Cloves	1300	800	300
Coriander	1200	300	200
Dill	1500	500	50
Geranium	1300	300	100
Lavender	1800	350	100
Lemon	1300	35	5
Limes	1200	25	3
Mandarin	1300	15	2
Neroli	1250	300	100
Orange	1200	300	2
Peppermint—			
Mitcham	950	300	80
French	1000	300	15
American	900	300	18
Japanese	1000	300	60
Petitgrain	1200	300	60
Pimento	3000	800	250
Rosemary	1500	350	150
Thyme	1000	450	100
Verbena	1100	300	100

required, as the difference in odour of various makes—of bergamot, for example, is most striking. The approximate quantities to be used when converting any formula can easily be arrived at by a reference to the table of concentrations given in Volume I, and these can be accurately adjusted by practice, until the exact reproduction of the odour with dilute alcohol becomes standardised. In order that the quantities of terpeneless oil for *different strengths* of alcohol may be quickly calculated, a table (after Gattefossé) is given on page 258.

To facilitate the conversion of a formula from the ordinary to the *terpeneless*, let us take a well-known example. The following was the prize-winner in a competition immortalised by the *Chemist and Druggist*.

The original was as follows:

No. 1168

Bergamot oil	2 fluid drachms
Lemon oil	1 fluid drachm
Neroli oil	20 minims
Origanum oil	6 minims
Rosemary oil	20 minims
Rectified spirit	1 pint
Orange flower water triple	1 fluid ounce

On changing the quantities of all our oils to minims and dividing by the concentrations 3, 16, 3, 3, and 4 our formula reads:

No. 1169

Bergamot oil	40 minims
Lemon oil	4 minims
Neroli oil	7 minims
Origanum oil	2 minims
Rosemary oil	5 minims
Alcohol, 60 per cent	1 pint
Orange flower water triple	1 fluid ounce

The product will not require filtration and will not deposit on standing. Further, we have as good a perfume at two-thirds of the cost!

Cheap Colognes

The use of industrial alcohols has reduced this problem to one of substituting expensive products with cheaper ones and lowering the percentage of essence contained to the limit that the public will accept. The use of iso-propyl alcohol offers no advantages because its odour is not so good. The following substitutions will be found effective:

Bergamot	Linalyl acetate
Lemon	Citral from lemon-grass

Sweet orange and French petitgrain	Petitgrain (Paraguay) or methyl anthranilate or aurantiol or one of the naphthol ethers
Neroli petales	Neroli, synthetic, or methyl naphthyl ketone
Rose otto	Rhodinol or geraniol
Clary sage	Linalyl propionate
Lavender	Terpinyl acetate
Coriander	Linalol
Musk extract	Musk ketone
Ambergris extract	Labdanum R.

The greater number of the other commonly used ingredients are not costly and therefore no substitution is necessary.

Two examples are given—No. 1170 being higher priced than No. 1171.

No. 1170

100	Bergamot
200	Linalyl acetate
200	Lemon
200	Sweet orange
100	Petitgrain, Paraguay
70	Neroli, synthetic
10	Rhodinol
40	Lavendin, 20/22 per cent esters
30	Rosemary, French
10	Cloves
10	Musk ketone
30	Benzoin R.
1000	

Dissolve 5 per cent or thereabouts in 80 per cent alcohol. Stand aside for one week and then filter through kaolin.

No. 1171

600	Linalyl acetate
200	Citral
50	Aurantiol
30	Ceylon citronella
50	Terpinyl acetate
50	Rosemary, Spanish
20	Musk xylene
1000	

Use 1 or 2 per cent in weak alcohol.

'Frozen' Eau de Cologne

'Frozen' Eau de Cologne is a novel and convenient form in which this useful and refreshing toilet article may be carried in the hand-bag. It is a solid and transparent alcoholic soap, prepared by dissolving 5 per cent of sodium stearate in warm alcohol, to which has been added the oils, and carefully running into suitable moulds. Another method preferred by some manufacturers is to dissolve at gentle heat a first-class milling base in alcohol direct. Ten per cent of soap chips or thereabout make a fairly satisfactory product. This is often slightly tinted with a small quantity of chlorophyll or other green dyestuff, and is then run into moulds, which are cooled until solidification is effected. The stick should be wrapped in tinfoil or cellophane to facilitate handling, and packed for sale in air-tight tubes. Exceptional cooling effects are obtained by the addition of suitable quantities of menthol.

Lavender water

This is undoubtedly England's most famous perfumed *eau de toilette*, and was originally prepared by macerating the fresh flowers in alcohol and then distilling the mixture. Nowadays the best products are made by maturing the blended ingredients for a prolonged period; if English lavender oil is used, it should be of at least one year's maturity, and the bouquet of the finished product is much improved by the addition of about one-half of the quantity of bergamot oil. The best fixatives are civet and musk, but they may be replaced by other aromatics chosen from the odour classification which follows.

A good formula for this type is as follows:

No. 1172

20	Lavender oil
10	Bergamot oil
5	Tincture of musk
5	Tincture of orris
1000	Alcohol, 90 per cent
1040	

Mature for *at least six* months and filter if necessary.

Odour classification

Top notes	Middle notes	Basic notes
1. Benzyl acetate	15. Cinnamyl formate	70. Linalyl salicylate
2. Linalol	Guaiac wood	73. Cassie absolute
4. Lavender	16. Serpolet	79. Civet absolute
Phenyl ethyl alcohol	17. Melissa	80. Hydroxy citronellal
5. Terpinyl acetate	18. Calamus Marjoram	90. Estragon
6. Bergamot	Orris absolute	Opoponax resin
Linalyl benzoate	20. Clary sage	Rhodinyl acetate
Phenyl ethyl benzoate	21. Angelica seed Ionone alpha	100. Ambergris Benzoin
7. Geraniol, Java	Myrrh resin	Castoreum absolute
8. Geraniol palmarosa	Rosemary	
Lemon	22. Clove	Costus
Linalyl propionate	Geranyl formate	Coumarin
Nerol	23. Parsley	Ethyl vanillin
Rhodinol	24. Benzylidene acetone	Labdanum resin
9. Neryl acetate		Musk ketone
Peppermint	Cinnamon bark	Oakmoss
10. Camomile	Ethyl cinnamate	Olibanum resin
Lavendin	Geranium grasse	Patchouli
Linalyl acetate	Rose otto	Santal
11. Linalyl cinnamate	Ylang	Tonka resin
Linalyl formate	30. Ambrette seed	Vetivert
12. Petitgrain, French	Cardamon	
Terpinyl propionate	31. Orange flower absolute	
13. Orris concrete	33. Zdravets	
	40. Hyssop	
	41. Mace	
	43. Jasmin absolute	
	Rose absolute	
	55. Nerolidol	

Owing to the high price of English lavender oil the judicious use of fine quality *French* oil may be necessary, and in the latter case ageing of the oil can be dispensed with.

No. 1173

5	Lavender oil, English
15	Lavender oil, French
10	Bergamot oil

1	Clary sage oil
4	Civet extract, 3 per cent
5	Orris infusion
1000	Alcohol, 90 per cent
1040	

Another more elaborate variation of this type is as follows:

No. 1174

4	Lavender oil, English
12	Lavender oil, French
0·5	Neroli oil
2	Bergamot oil
1	Sweet orange oil
5	Musk extract, 3 per cent
5	Civet extract, 3 per cent
3	Styrax infusion
2	Orris tincture
2	Ambrette oil, 10 per cent
1	Vanillin
0·5	Benzyl iso-eugenol
1000	Alcohol, 90 per cent
1038	

Quickly matured lavenders

Quickly matured lavenders are prepared by mixing the oils for a few weeks and diluting with alcohol before use. The modern tendency is to fix with vanilla and oakmoss as well as the animal extracts. Coumarin makes a useful addition as follows:

No. 1175

600	Lavender oil, French
250	Bergamot oil
20	Neroli, compound
10	Rose rouge, compound
20	Orris, concrete
50	Vanilla extract, 10 per cent
20	Civet extract, 3 per cent
20	Musk extract, 3 per cent
1	Oakmoss absolute
1	Patchouli oil
8	Coumarin
1000	

Dissolve 7 to 10 per cent in 80 per cent alcohol and after one week freeze and filter bright.

Amber lavenders

Some people have a distinct preference for this heavier type of lavender perfume, and the fineness of the bouquet produced by the English oil is much impaired by the presence of so much civet, etc. A really good French oil is therefore all that is necessary to account for the lavender odour in the perfume. Two formulae follow:

No. 1176

25	Lavender oil, French
5	Bergamot oil
10	Civet extract, 3 per cent
5	Musk extract, 3 per cent
2	Coumarin
1	Vanillin
2	Benzoin resin
1000	Alcohol, 90 per cent
1050	

No. 1177

20	Lavender oil, French
10	Bergamot oil
2	Lemon oil
3	Clary sage oil
1	Musk ambrette
5	Castoreum extract, 3 per cent
25	Essence ambergris, 3 per cent
1000	Alcohol, 80 per cent
1066	

Other types of lavender water are represented by the following formula, where the introduction of ylang-ylang and jasmin adds a new flower note, and this, together with the slightly camphoraceous odour of the rosemary, produces a very pleasant variation:

No. 1178

25	Lavender oil, French
5	Rose, compound
1	Ylang-ylang oil, Manila

1	Jasmin absolute
2	Rosemary oil
0·25	Nutmeg oil
1	Oleo-resin orris
900	Alcohol, 90 per cent
100	Orange flower water triple
1035	

Add the orange flower water after the other ingredients have macerated one month. Filter bright if necessary.

Terpeneless products are easily prepared from any formula by calculating the different quantities of oils that will be required (*see* table in Volume I). The following preparation is a good one:

No. 1179

10	Lavender oil, French, terpeneless
4	Bergamot oil, terpeneless
10	Essence of musk, 3 per cent
1	Coumarin
1000	Alcohol, 60 per cent
1025	

Macerate with frequent shaking for six months.

Cheap lavender perfumes are made by substitution of some of the more expensive ingredients, as follows:

No. 1180

600	Lavendin 20/22 per cent esters
250	Linalyl acetate
20	Aurantiol
30	Geranyl acetate
10	Ionone, 100 per cent
40	Benzoin R.
20	Coumarin
30	Musk xylene
1000	

Dissolve 1 or 2 per cent in weak alcohol.

Florida water

Florida water is a popular perfume in America and a native of that country, just as lavender water is of England. It very closely resembles a mixture of eau de Cologne and lavender water to which has been added cinnamic aldehyde and eugenol. The following formulae will indicate the lines upon which it may be prepared:

No. 1181

15	Lavender oil, French
5	Portugal oil
25	Bergamot oil
10	Petitgrain oil, Paraguay
1	Eugenol
1	Cinnamic aldehyde
5	Rose geranium oil
2	Oleo-resin orris
1	Musk ambrette
200	Orange flower water triple
800	Alcohol, 90 per cent
1065	

A better product is obtained as follows:

No. 1182

5	Neroli oil, bigarade
7	Lavender oil, French
30	Bergamot oil
2	Limes oil
2	Clove oil
1	Cassia oil
1	Cinnamon oil
5	Rose otto, synthetic
2	Amber, synthetic
100	Orange flower water triple
900	Alcohol, 90 per cent
1055	

Eau de Cananga approximates very closely to eau de Cologne, and most formulae only show a difference in that the neroli is replaced by cananga or ylang-ylang:

No. 1183

5	Cananga oil
4	Bergamot oil
1	Lemon oil

TOILET WATERS: FLORIDA WATER

0·2	Benzaldehyde
2	Musk extract, 3 per cent
1000	Alcohol, 80 per cent
1012	

Eau de Portugal also resembles eau de Cologne, *sine neroli*:

No. 1184

20	Sweet orange oil
6	Bergamot oil
3	Lemon oil
1	Rose geranium oil
2	Benzoin R.
1000	Alcohol, 90 per cent
1032	

CHAPTER EIGHT

Soap Perfumery

Soap manufacture

The manufacture of toilet soap may be divided broadly into two stages:

1. The production of the soap base by saponification of fats, oils, etc.
2. The perfuming, milling, plodding, and stamping of this base yielding the finished tablets of toilet soap.

The object of this chapter will be to treat very briefly the first stage and to go into details concerning the second stage. The perfuming of toilet soap is an art acquired by long practice, keen observation of the relationship existing between each aromatic substance and the soap, and attention to small details in the course of milling, plodding, etc. While some cheap soaps are perfumed in the course of the manufacture of the base, all the finer products exhibited on the shop counters are perfumed afterwards, and today constitutes, in many cases, a separate business.

There are several qualities of milling chips, some firms manufacturing as many as five. These vary in price according to the raw materials used, and differ widely in colour from an almost pure white to a greyish-yellowish-brown. Practically all these bases are made by boiling. The cold process soaps are seldom an unqualified success since they are rarely neutral. Small makers sometimes manufacture them owing to their easier production.

Raw materials

In the manufacture of the best quality soap it is necessary to employ the purest, whitest, and least odorous fats and oils. Those of importance are tallow, cochin coconut oil, castor oil, lard, and palm kernel oil. Caustic soda is the alkali almost invariably employed for saponification. This makes a hard and stable soap. Potash produces a soft article. The price and quality of the soap stock varies very much according to the proportion of tallow employed. When this is replaced by hardened oils, a higher proportion of coconut oil is used, together with some resin, which makes the finished soap sufficiently plastic for milling. The use of palm kernel oil gives the chips a creamy colour. Darker bases owe their colour to the use of cheaper oils, such as ground nut, etc.

Best quality soap stock can be made from the following:

	per cent
Finest white tallow	50-70
Cochin coconut oil	25-30
Lard	up to 20
Palm kernel oil	up to 20
Castor oil	up to 5
Resin	traces

The percentages of oils and soft fats are regulated according to the melting-point of the tallow. Castor oil improves the finish. Rancid fats are never used in good quality soaps. They may be made usable by saponifying the free fatty acids and completely separating the odorous soap from the clear fat. This can be done successfully with poor quality vegetable oils, margarines, and coconut oil.

In the larger works the fats are melted by steam blown into the barrels and the liquefied contents pumped to the storage tanks, being run into the pans as required. The caustic soda is generally made by the action of lime on a boiling solution of sodium carbonate. It is evaporated to the necessary specific gravity (degrees Baumé) before being run into the pans.

The boiling process

The boiling process may be divided into four stages:

1. *Pasting*, or the preliminary admixture of weak lye and melted fats and oils.

2. *Graining*, or the separation or salting out of the soap.

3. *Boiling on strength* (the most important part of the process), or the further addition of caustic lye from time to time throughout the day to keep the soap 'open'. This results in the removal of salt and impurities.

4. *Separation* (involving up to a week), or the dissociation of scum, pure soap about 80 per cent, nigre about 20 per cent, and half-spent lyes.

The half-spent lyes and nigre are run off for use again and the pure soap is then made into chips. All modern processes are based on the original Cressonières fundamentals, whereby the liquid soap on leaving the pan is run over cooled rollers and the semi-solid ribbons conveyed by bands through a drying chamber. This takes from 20 to 30 minutes and eliminates about 18 per cent of moisture.

The dry chips contain from 75 to 80 per cent of fatty acids and from 8 to 15 per cent of water. The quantity of sodium chloride should be less than 0·25 per cent since a higher percentage will cause the soap to crack and split badly after milling and plodding. The limit of alkalinity is about 0·006 per cent of free soda calculated as sodium oxide.

The cold process consists in the saponification of coconut oil at low temperatures with high strength alkali. Sometimes small quantities of other fats and oils are added to the coconut oil, and for some superfatted soaps lanolin is added at this stage. The exact quantity of alkali necessary for complete saponification is worked out previously, so that a neutral soap results. In making these soaps it is customary to melt part of the fats, remove the source of heat, and then add the remainder so that when all is liquid the temperature is between 30° and 40°C. The lye at about 40° Baume is run in gradually while the whole is gently agitated. As saponification takes place the mass becomes uniform and finally translucent. The perfume is now added, and after stirring, the whole is run into frames. Complete saponification is here effected when the temperature rises to about 85°C.

Shaving soap

Shaving soap differs from toilet soap essentially in that saponification is generally effected with potash as well as soda lye, and further that absolute neutrality must be obtained in the finished article to prevent irritation. A creamy and lasting lather is also important. The raw materials must be of the finest and consist principally of tallow and cochin coconut oil, together with small quantities of lard oil, castor oil, and lanolin. Soothing and emollient properties are improved by the addition, during milling, of small percentages of soft white paraffin, tragacanth (as mucilage), and glycerine. It is no use boiling the soap with a mixture of soda and potash lyes since double decomposition follows during graining. This difficulty may be overcome by saponification with soda as indicated above, and after separation of the pure soap, and while it is still liquid, crutching in with neutral potassium stearate: the latter being made by the direct saponification of stearic acid. The soap is run from the crutcher into frames to cool. It is subsequently chipped and milled as indicated below.

Transparent soaps

Transparent soaps may be prepared in several ways. One of the oldest methods is to dissolve a good quality soap in alcohol by the aid of gentle heat and to then distil off about 80 per cent of the alcohol and run the transparent liquid soap into moulds. This method is expensive but good. Ways of economising have been found by making additions of sugar, castor oil, and glycerine during the process for ordinary good quality millings. In the former method of manufacture a necessary prerequisite for success is a first-class milling base. Such a raw material is more easily and satisfactorily converted from the crystalline to the colloidal state which is really the essence of the process. There is of course no necessity to use duty-paid ethyl alcohol, a good quality industrial spirit, pure methyl alcohol or even iso-propyl alcohol yield good results. The moisture content of the milling chips is important, for the higher this is, the less transparent and brilliant is the finished tablet. Drying should be conducted at a fairly low temperature so that a good colour is preserved. The moisture content should at the completion of this process not exceed 5 per cent. For each hundredweight of soap a 20 gallon jacketed still having stirring apparatus and bottom and side exits is necessary. It is preferable to employ two, one for solution of the soap in the alcohol and the other for recovery of the latter. The recovery kettle should be shallow to allow a large evaporating surface. The temperature is usually about 75°C. The quantity of alcohol necessary for solution varies according to the soap used and to the skill of the operator; 1 hundredweight of chips will in one case require 5 or 6 gallons of spirit while in another as much as 10 gallons will be necessary. Settling of the dissolved soap demands careful attention and may require up to six hours. The clear supernatant solution is run off into the recovery kettle from the side exit and the residues withdrawn from the bottom for further treatment. The addition of glycerine at this stage in the process is general. The quantity should be kept low owing to its hygroscopic nature; 1 or 2 per cent is sufficient and will aid transparency, more than this will produce tackiness and a dull finish. The temperature is now maintained at 75°C and sometimes slightly increased until 75 or 80 per cent of the added alcohol has been recovered. This depends upon how firm the soap will set, and experimental tests are made from time to time to ensure success. The kettle is allowed to cool and then the perfume and any colour in clear solution are added. The liquid soap is now run out into narrow frames or bar moulds and subsequently cut to the size necessary for stamping. Shrinkage will of course occur as the last traces of alcohol evaporate, but this is generally allowed for before cutting. A skilled operator will have adjusted his process so that brilliant transparency and hardness are secured within a few days of manipulation.

Some years ago a patent was taken out by P. Villain for the preparation of transparent soap having a high perfume content by the use of a soluble form or derivative of cellulose. This is alleged to form a film coating on the surface of the soap which prevents deterioration of the perfume.

When buying soap chips it is usually advisable to send the contract samples to a competent analyst for report. The following physical characteristics should also be noticed: colour, dampness, odour, and taste. The first quality chips are generally white, the second slightly creamy, the third darker, and the fourth and fifth brownish. For an average good sample of toilet soap a mixture of equal parts of Nos. 1 and 2 are excellent. Many firms, however, rely entirely on No. 2 for first quality toilets and a mixture in equal parts of 2 and 3 for lower grades. A damp base is more difficult to mill than a normal sample. It is better to buy on the dry side and add the necessary quantity of water during mixing and prior to milling. The first quality chips are practically odourless, the second and third grades have a slight odour reminiscent of traces of para-cresol methyl ether. The lower grades have sometimes quite a strong smell. They can only be used in coloured soaps, but owing to the larger quantity of perfume necessary to cover the odour it generally pays in the long run to buy a better grade base and use less perfume. If the soap tastes salty it should be discarded, since it will crack after plodding and soon become unsaleable. When an analyst's report is obtained the fatty acid content and percentage of free alkali or free fat should be noted. Absolute neutrality is very seldom found, but it is better to have traces of free alkali than traces of free fat. The latter soon turn the soap rancid and the quantity of perfume used becomes prohibitive. Free alkali is prone to affect the perfume by the decomposition of esters, etc. This may be counteracted to some extent by the addition of lanoline (about 1 per cent) and the use in the perfume of gum resins such as storax. A dark coloured base can be much improved by the addition of from $\frac{1}{2}$ to 1 per cent of zinc oxide or titanium dioxide.

Super-fatted toilet soaps
Super-fatted toilet soaps are made by the addition of from 1 to 5 per cent of the following during the milling process:

    Anhydrous lanolin   Soft paraffin jelly

In both cases the substances are liquefied before being added to the soap chips in the mixer. In the former it is often stirred with a little hot water beforehand. The finished soap tablet has a more shiny appearance and a velvety softness in use. The mere traces of lanolin or soft paraffin left on the skin after washing have a distinctly beneficial effect.

The milling process
The milling process is preceded by mixing. This is effected in a galvanised iron apparatus in which a bent arm revolves. The soap chips, perfume, colour, and superfat are placed together and the arm allowed to revolve until uniform mixation is completed. The apparatus is tipped over and the contents transferred to the hopper of the milling machine. This consists generally of four

granite[1] rollers so arranged that as the ribbons come from the top roller they can be dropped when desired on to the two bottom rollers again. The speed at which these four rollers revolve is so regulated, each faster than the other, that the thin sheet of milled soap passes from the bottom to the top one automatically. Four millings are generally enough for all purposes—too much milling results in transparency. After the fourth milling the ribbons are run off the rollers into suitable containers, from which the contents may be easily transferred to the plodder. This is constructed very much on the lines of a huge sausage machine, having a hopper, central screw, and conical-shaped nozzle. This is heated to give a glossy finish to the soap. The temperature required varies with different bases. It is generally between 40° and 60°C. The screw must be well covered with ribbons to ensure perfect compression. The soap comes from the nozzle in a bar, the size of which is adjusted by the insertion of different plates. The aperture must be kept perfectly smooth (with emery paper) to prevent striation of the soap bar. This is cut to suitable lengths by wire and subsequently to size of tablet in another cutting machine. In large scale production, however, the tablets are automatically cut to size as the bar is forced from the nozzle of the plodder. The tablets are now stamped to the desired shape in machines having moulds made of stainless steel. They are stood aside on trays for a few hours and rough edges trimmed off with knives or scrapers.

Coloured soaps

These can be improved, as indicated above, by the use of zinc oxide or titanium dioxide. This is also useful for stabilising the colour in coloured soaps. Generally dyestuffs are to be preferred to pigments.

Dyestuffs require very careful testing before use. They must be perfectly soluble in hot water, not cause spottiness in the soap, be stable when exposed to light and not used in sufficient quantity to dye any garments. To test the latter it is sufficient that the froth caused by the coloured soap should be white. In general most colours are stable. The important exception is violet, but this difficulty can generally be overcome by blending reds and blues.

Cracking in toilet soaps is a troublesome feature which crops up

---

[1] Nowadays water-cooled steel rollers are more common. The output per mill is much greater.

from time to time in the majority of soap plants. It is often attributed to either uneven pressure in the plodder or the presence of too much salt in the soap chips. While it is perfectly true that either of these faults may be responsible for cracked soap, it does not necessarily follow that they are the only cause, because experienced soapmakers are able to control the salt content to a fine point and the man in charge of the milling plant keeps a close eye on the plodder with a view to preventing any insufficient feeding with chips. Readers will have noticed that cracking in soaps is most prevalent from December to March, especially when these months are very cold. During milling the temperature of the soap rises from 10° to 15°C, and it is frequently allowed to stand about afterwards in bins in a cold atmosphere for sufficient time for the temperature to fall again. It is then fed into the plodder where the temperature again rises, and after the stamped tablets have cooled, cracks appear. The author has found that this can be prevented by keeping the temperature of the chips about 30°C between mill and plodder.

Perfumes

The cost of a toilet soap is determined largely by the perfume. A perfect sample should be strong without being harsh and should retain its fragrance right up to the last thin wafer in use. The stability in the presence of alkali of the aromatic substances in any compound is of importance, as also any change in colour which they may effect in the soap after manufacture. Successful soap perfuming therefore requires a knowledge of the durability of each individual raw material when in contact with the soap and any of its constituent impurities. One or 2 per cent of compounded perfume, spread through a large and nonvolatile mass has to yield an odour suggestive of the finer alcoholic perfumes in which the solvent plays an important part in determining the delicacy of the finished odour. While alcohol develops the finer ingredients of a perfume, soap on the contrary modifies the odour to such an extent as to often make it unrecognisable in the absence of correct blending. Raw materials listed in the Odour Classification under Basic Notes play an important part in solving this problem, and the durability of a perfume in soap largely depends upon their correct selection. Natural oils and resins are often the key, and if price will allow they should be used liberally. Some makers rely upon a very small range of raw materials, and in consequence their products lack variety. Many synthetics have proved of great value in soap perfumery, but in all cases, before adoption, experiment is necessary. For this purpose a miniature mill and plodder are made by a well-known firm. To save expense, however, many chemists use an ordinary mincing machine, and after passing the perfumed ribbons through two or three times, pound them into a mass in a mortar, and stand aside under ordinary commercial conditions for three months. At the end of this time the durability of any perfume compound can be well judged. To facilitate choice a list of substances with their properties is appended. The stability and keeping qualities are indicated as follows—very good, A; good, B; weak, C; poor, D. These are all that are necessary for the standard lines of soap, but may be added to when the creation of a new bouquet is desired:

Ajowan oil will replace thyme oil. It is stronger and imparts a pleasing freshness.   A.

Almond oil, bitter, useful in acacia and fancy compounds, generally replaced by benzaldehyde, free from chlorine.   C.

Ambergris tincture, not often used excepting in very high class soaps, generally replaced by labdanum.   B.

Amyl cinnamic aldehyde, useful in jasmin soaps and very persistent.   B.

Amyl salicylate, very good in trèfle compounds and quite strong and lasting in use.   A.

Aniseed oil, very powerful. Use traces only.   B.

Anisic aldehyde, excellent in May blossom and acacia compounds. About 10 per cent sufficient.   A.

Atlas cedarwood oil, excellent in acacia, mimosa, santal, and violet. Very strong and lasting. Resinoid also good in fern and chypre.   A.

Basil oil, indispensable for mignonette, but requires to be well blended.   B.

Bay oil, sometimes used in place of clove oil.   B.

Benzaldehyde, cheap and useful in almond soaps, traces in fancy compounds yield pleasing results.   C.

Benzoin, used as a resinoid and also as a tincture made with industrial alcohol, inclined to darken soaps.   B.

Benzyl acetate, 5 or 10 per cent will give a refreshing and sweet odour to most soaps, useful in jasmin and orange blossom when blended with petitgrain oil.   B.

Benzyl alcohol, weak and faintly balsamic, useful as a diluent.   C.

Benzyl benzoate, a good solvent for musk xylene.   A.

Benzylidene acetone, useful in the sweet-pea type, blends well with bromstyrole for cheap bouquets. Owing to its irritating effect on the skin, should be used with care.   B.

Bergamot oil, very good, often replaced by lemon oil.   A.

Beta naphthol ethers, very powerful, but require to be well blended, otherwise coarse. Use in moderation.   A.

Bois de rose oil, indispensable in all flower compounds, much improves Bourbon geranium.   A.

Bornyl acetate, the base of all pine perfumes, useful in lavender compounds.   A.

Bromstyrole, very powerful and stable, good hyacinth base.   A.

Cananga oil, useful in violet and santalwood, not very strong and lacks body.   D.

Caraway oil, an indispensable constituent in brown windsor. Rather fleeting.   B.

Cassia oil, base for all brown windsors, good blender in bouquets, and should not be overlooked in some rose compounds. Causes darkening.   A.

Castoreum tincture, excellent and cheap substitute for musk.   A.

Cedarwood oil, strong and persistent, good basis for violets.   A.

Cinnamic alcohol, weak but good hyacinth base.   C.

Cinnamic aldehyde, occasionally used to replace cassia oil.   A.

Cinnamon leaf oil, good substitute for clove oil, darkens in soap.   C.

Citral, sometimes used as a substitute for lemon oil.   B.

Citronella oils. Both Java and Ceylon very largely employed in all forms of cheap soap compounds, Ceylon frequently preferred and not always on account of its lower cost—very strong.   A.

Citronellol, useful in rose compounds.   B.

Citronellyl esters, useful as modifiers and blenders.   C.

Civet tincture, made with industrial spirit, often replaced by synthetic civets based on phenylacetic acid and skatole.   A.

Clove oil, much used for imparting a pleasant sweetness to compounds and indispensable in carnation, darkens rapidly. A.

Copaiba resin (balsam), good fixative. A.

Coumarin, very good in all compounds, not forgetting lavender. Gives a yellowish tinge in time. A.

Para-cresol methyl ether, very strong and stable. Use only about 1 per cent. A.

Diphenyl methane and oxide, very strong and stable, substitutes for geranium oil. A.

Ethyl cinnamate, very powerful and balsamic. B.

Eucalyptus oil, traces useful in fern compounds, much used in medicated soaps. A.

Eugenol, good in carnation, but darkens less than clove. B.

Fennel oil, very useful, requires moderation. C.

Geraniol, stable in rose and other compounds. A.

Geranium oils, both Algerian and Bourbon indispensable, latter rather coarse, equal parts of each give best results, used in rose and all bouquets. A.

Geranyl acetate, excellent in rose and lavender. C.

Ginger-grass oil, very good, but requires to be well blended. B.

Guaiac-wood oil, an excellent base for all compounds. A.

Heliotropin, only permissible in coloured soaps, odour good, and indispensable in heliotropes. A.

Hydroxy-citronellal. Use the residues—quite stable and persistent. C.

Ionone, unpurified 100 per cent or residues will yield good results, about 20 per cent sufficient in violets, less in bouquets. B.

Iso-butyl esters, good modifiers and blenders, rather weak excepting the phenylacetate and salicylate. B.

Iso-eugenol, excellent in carnation but darkens rapidly, colours the soaps containing it. B.

Labdanum, very persistent in small quantities, will replace ambergris, good in lavender. A.

Lavender oil, useless alone, must be strengthened and well blended, oakmoss excellent for the purpose, also rosemary, thyme, and borneol. B.

Lavender oil, spike, powerful and very good in lavenders, ferns, etc. A.

Lemon-grass oil, basis of all verbenas. Modify with bois de rose or palmarosa—turns the soap yellowish. C.

Lemon oil, inclined to darken slightly. Ten per cent will make a marked difference to many compounds, indispensable in Cologne soaps, but must be well fixed to retard oxidation of terpenes.   C.

Linalol. Better use bois de rose oil.   A.

Linalyl acetate, good, but better to use bergamot oil.   B.

Methyl acetophenone, must be used in small quantities and requires well blending, useful in acacias and mimosas.   A.

Methyl anthranilate, very cheap and strong, but inclined to darken.   A.

Methyl benzoate, very strong and stable.   A.

Methyl cinnamate, better than ethyl ester, small quantities give an amber note.   A.

Methyl heptine carbonate, traces sometimes used in violets, sharp at first but softens after a few days.   C.

Methyl salicylate, used in chypre and fern soaps.   B.

Mirbane oil. Do not use it.   B.

Musk ambrette, about 2 per cent is excellent, more gives a sickly sweetness. Residues are good and cheaper.   A.

Musk tincture, seldom employed.   B.

Musk xylene, very useful, but make sure it is all dissolved, otherwise the soap will be spotty—turns yellowish in time.   A.

Nutmeg oil, used occasionally in lavender compounds.   D.

Oakmoss, excellent in lavender and fern, much used in fancy bouquets, very persistent. The green resin is good enough for all purposes.   A.

Orange oil, excellent sweetener but must be well fixed, inclined to darken.   C.

Orris oil concrete, much used in violets, but often replaced with the oleo-resin or resin extracted from the spent rhizomes. Industrial tinctures of this material useful.   B.

Palmarosa oil, good and stable, useful in rose and similar compounds.   A.

Patchouli oil, very good and stable, more than 10 per cent becomes unpleasant, much less generally enough, useless in rancid bases.   A.

Pepper oil, traces employed in fancy bouquets and carnation compounds.   B.

Peppermint oil, traces are useful for developing other odours.   A.

Peru balsam, imparts warmth, darkens in time.   A.

Petitgrain oil, good in many compounds, but turns yellowish in time, often used for predominating note in glycerine and cucumber and Cologne soaps. Very durable. B.

Phenylacetic aldehyde, not very valuable in soap, better use bromstyrole. D.

Phenylethyl alcohol, useful in all flower compounds. A.

Phenylethyl esters, good blenders, rather weak. B.

Rhodinol, excellent in rose and other compounds, usually too expensive. Use Bourbon Geranium. A.

Rosemary oil, good and stable, imparts freshness, useful in many compounds, especially lavender. A.

Safrole, much used in cheap soaps. A.

Sandalwood oil, excellent blender in violets and roses, requires to be developed in santal soaps. A.

Sassafras oil, indispensable in chypre. A.

Spearmint, powerful at first but evanescent. C.

Storax, the best all-round resinous fixative. A.

Terpineol, often the only perfume in cheap soaps, excels rather as a basis on which to build fanciful odours. Use 10 per cent or thereabouts—very stable and strong. A.

Terpinyl esters, fair blenders, but not so stable. B.

Thyme oil, excellent for imparting freshness to all kinds of soap compounds, antiseptic value good. A.

Tolu balsam, good fixative, but inclined to darken. B.

Vanillin, traces useful in fancy bouquets, turns the soap yellowish. B.

Vetivert oil, small quantities are most persistent, good in violets. A.

Flower oils are only used occasionally in the highest priced soaps. They yield excellent results. A cheaper jasmin is made at Grasse by distilling the flowers with cedarwood oil. This product is of considerable value. Mimosa absolute should not be overlooked since it is fairly cheap.

Matching a perfume of known composition requires experiment and is often necessary when a toilet soap is included in any series of cosmetic products. As will be readily understood it would be impossible, from a pecuniary point of view, to use a first-class compound in soaps. From another point of view it would be equally unsatisfactory, since some of the ingredients might undergo decomposition, cause discoloration, or generally upset the finished odour to such an extent as to make it unrecognisable. A formula therefore requires very careful examination and analysis. Those ingredients which are known to be stable and not too dear are allowed to remain although the quantities may have to be adjusted. The others are eliminated but replaced with raw materials having similar odours. Other substances may have to be added to diffuse the perfume. The most important point is to choose the right combination of basic notes and add just sufficient to regulate the volatility of the perfume. There are, of course, many essential oils having distinctive odours which at the same time act in this way as for instance, vetivert and santal in violets; patchouli and guaiac-wood in roses. These are often more stable than synthetics but, nevertheless, gum-resins in liberal quantity are imperative constituents. The soap perfumer's stand-by is styrax, but this is by no means the only one of value; almost all the resinous substances are excellent and many useful variations can be made with them. It should be noted that it is useless to judge the odour of a soap compound by the usual methods. *The perfume must be incorporated in the soap by milling and the finished tablet left on one side for three months.* As stated above, the experiments leading up to this stage may be carried out with a mincing machine, etc.

In order to indicate the lines upon which matching may be conducted an example of violet will be examined. A good quality compound for general use in perfumery will be approximately as follows:

No. 1185

500	Ionone alpha (1)
80	Orris oil, concrete (2)
50	Heliotropin (3)
150	Bergamot oil (4)
30	Ylang-ylang oil, Manila (5)
40	Sandalwood oil (6)
30	Violet leaf absolute (7)
25	Cassie absolute (8)
50	Jasmin absolute (9)
1	Aldehyde C_{12} (10)
25	Rose otto, Bulgarian (11)
20	Benzoin R. (12)
1001	

This formula would be adjusted as follows:

Replace 1 with ionone 100 per cent for soaps and reduce the quantity: Replace 2 with orris oleo-resin and reduce, only when costs compel: Retain 3 and 4: Replace 5 with cananga oil and increase: Increase 6: Replace 7 with methyl heptine carbonate and reduce: Omit 8: Replace 9 with benzyl acetate or inexpensive jasmin compound: Omit 10: Replace 11 with geranium oil and increase: Retain 12: To strengthen and diffuse add cedarwood, clove, and vetivert oils, terpineol and musk xylene or residues.

The formula for the violet soap compound having a similar odour to the above would therefore read:

No. 1186

200	Ionone, 100 per cent for soaps
80	Orris oleo-resin
50	Heliotropin
150	Bergamot oil
50	Cananga oil
60	Sandalwood oil
1	Methyl heptine carbonate
50	Benzyl acetate
40	Geranium oil
20	Benzoin R.
130	Cedarwood oil
20	Clove oil
30	Vetivert oil
50	Musk residues
70	Terpineol
1001	

In making violet soaps some manufacturers add palm millings which emit a violet-like odour. Others add powdered orris-roots, but the use of the oleo resin dispenses with the necessity.

Soap compounds

Two examples of each of the principal toilet soap perfumes will now be given. The first and better quality one being based where possible largely upon essential oils, and the second and cheaper more particularly upon synthetics. They are all workable and are capable of endless modification to suit individual tastes. By applying the principles outlined above it will be quite easy to further cheapen the compound as desired. With a view to assisting the experimenter in the choice of raw materials for odours of specific type, the more important components are enumerated in accordance with my classification of odours.

ACACIA

Odour classification

Top notes	Middle notes	Basic notes
1. Benzyl acetate	15. Acet anisol	77. Methyl naphthyl ketone
2. Rosewood	21. Anisic aldehyde	
3. Methyl acetophenone	Ionone	80. Hydroxy citronellal
Petitgrain para	Methyl anthranilate	91. Undecalactone
Terpineol	22. Clove	100. Benzoin resin
4. Bromstyrole	24. Geranium	Coumarin
Phenyl ethyl alcohol		Labdanum resin
6. Bergamot		Musk xylene
Iso-butyl phenylacetate		Peru balsam
		Phenyl acetic acid
7. Geraniol, Java		Santal
8. Cedarwood		Styrax resin
Iso-butyl benzoate		Tolu balsam
10. Diphenyl oxide		Vetivert
Linalyl acetate		
12. Methyl cinnamate		
14. Cananga		

Acacia Soap, no. 1187

100	Rosewood
20	Methyl acetophenone
200	Petitgrain para
50	Terpineol
50	Bergamot
20	Methyl cinnamate
40	Cananga
200	Anisic aldehyde
50	Methyl anthranilate
20	Clove
50	Geranium, African
50	Methyl naphthyl ketone
50	Santal
30	Tolu balsam
40	Vetivert
30	Musk xylene
1000	

Acacia Soap, no. 1188

50	Rosewood
100	Petitgrain para
20	Bromstyrole
100	Terpineol
100	Geraniol, Java
100	Cedarwood
20	Methyl cinnamate
40	Cananga
200	Anisic aldehyde
50	Ionone
50	Methyl anthranilate
20	Clove
50	Methyl naphthyl ketone
70	Styrax resin
30	Musk xylene
1000	

ALMOND

Almond Soap, no. 1189

400	Benzaldehyde
150	Rosewood
100	Clove
50	Geranium, African
70	Orris resin
100	Peru balsam
100	Santal
30	Musk xylene
1000	

Odour classification

Top notes	Middle notes	Basic notes
1. Benzaldehyde	15. Acet anisol	90. Orris resin
Benzyl acetate	21. Anisic aldehyde	100. Benzoin resin
2. Rosewood	Ionone	Coumarin
6. Bergamot	Methyl anthranilate	Musk xylene
7. Geraniol, Java	22. Clove	Patchouli
8. Lemon	24. Geranium	Peru balsam
11. Sweet orange	45. Cassia	Santal
14. Cananga		Styrax resin
Fennel		Tolu balsam
		Vetivert

Almond Soap, no. 1190

500	Benzaldehyde
100	Rosewood
100	Geraniol, Java
50	Anisic aldehyde
100	Ionone
50	Cassia
70	Styrax resin
30	Musk xylene
1000	

AMBER

Amber Soap, no. 1191

150	Rosewood
50	Phenyl ethyl alcohol
150	Bergamot
100	Amyl salicylate
40	Geranium, Bourbon
10	Castoreum absolute
100	Coumarin
200	Labdanum resin
30	Musk xylene
20	Oakmoss
50	Phenyl acetic acid
100	Styrax resin
1000	

Odour classification

Top notes	Middle notes	Basic notes
1. Benzyl acetate	21. Ionone	100. Castoreum absolute
2. Rosewood	29. Geranium, Bourbon	Coumarin
4. Phenyl ethyl alcohol	42. Amyl cinnamate	Labdanum resin
6. Bergamot		Musk xylene
7. Geraniol, Java		Oakmoss
8. Amyl salicylate		Olibanum resin
Lemon		Patchouli
11. Sweet orange		Phenyl acetic acid
		Santal
		Styrax resin
		Vanillin
		Vetivert

Amber Soap, no. 1192

100	Benzyl acetate
100	Rosewood
100	Phenyl ethyl alcohol
100	Geraniol, Java
150	Lemon
100	Coumarin
100	Labdanum resin
50	Olibanum resin
50	Musk xylene
20	Patchouli
50	Santal
30	Vanillin
50	Vetivert
1000	

Buttermilk, no. 1193

100	Geraniol
200	Geranium, Bourbon
100	Clove
100	Terpineol
200	Cedarwood
100	Santal
80	Petitgrain
40	Patchouli
60	Benzyl acetate
20	Styrax resin
1000	

Buttermilk, no. 1194

200	Citronella, Java
100	Cedarwood
200	Guaiac-wood
100	Rosewood
200	Terpineol
30	Diphenyl methane
30	Musk xylene
100	Benzyl acetate
40	Peru balsam
1000	

Bouquet, no. 1195

200	Guaiac-wood
200	Rosewood
100	Geranium
10	Vetivert
20	Ionone residues
100	Cedarwood
100	Terpineol
90	Clove
100	Lavender
20	Oakmoss
30	Benzoin resin
30	Musk xylene
1000	

Bouquet, no. 1196

200	Terpineol
200	Rosewood
100	Lavender spike
30	Diphenyl oxide
100	Cedarwood
200	Citronella, Java
50	Clove
50	Linalyl acetate
20	Red thyme
20	Styrax resin
30	Musk xylene
1000	

BROWN WINDSOR

Odour classification

Top notes	Middle notes	Basic notes
3. Petitgrain para Terpineol	16. Eugenol	100. Castoreum absolute
4. Lavender	21. Ionone Rosemary	Iso-eugenol
6. Bay Bergamot Caraway	22. Cinnamon leaf Clove	Musk xylene Pimento
7. Geraniol, Java	45. Cassia	Styrax resin
8. Cedarwood Citronella, Ceylon		
9. Spike lavender Thyme, red		

Brown Windsor, no. 1197

200	Cassia
50	Caraway
100	Clove
100	Lavender
100	Rosemary
50	Thyme, red
100	Petitgrain
100	Bergamot
150	Cedarwood
20	Styrax resin
30	Musk xylene
1000	

Brown Windsor, no. 1198

100	Cassia
400	Citronella, Ceylon
100	Clove residues
200	Terpineol
30	Red thyme
100	Cedarwood oil
20	Caraway
50	Rosemary
1000	

Cedarwood, no. 1199

150	Santal
10	Bromstyrole
100	Cananga
100	Terpineol
150	Methyl ionone
250	Cedarwood
100	Geranium, Bourbon
30	Vetivert
60	Orris resin
50	Heliotropin
1000	

Cedar, no. 1200

600	Cedarwood
100	Santal
30	Ionone residues
100	Geraniol
100	Phenylethyl alcohol
20	Vetivert
50	Coumarin
1000	

CARNATION

Odour classification

Top notes	Middle notes	Basic notes
1. Benzaldehyde	15. Heliotropin	90. Orris resin
Benzyl acetate	16. Eugenol	100. Benzoin resin
2. Rosewood	21. Ionone	Iso-eugenol
3. Terpineol	22. Benzyl iso-eugenol	Musk xylene
4. Bromstyrole	Clove	Pepper
Phenyl ethyl alcohol	24. Cinnamon leaf	Peru balsam
	Geranium, African	Pimento
6. Bay	45. Cassia	Styrax resin
Bergamot		
8. Amyl salicylate		
Benzyl salicylate		
Geraniol palmarosa		
14. Cananga		

Carnation Soap, no. 1201

100	Amyl salicylate
100	Geraniol, Java
450	Eugenol
50	Geranium, Bourbon
30	Cassia
50	Orris resin
70	Iso-eugenol
30	Musk xylene
20	Pepper
100	Peru balsam
1000	

Carnation Soap, no. 1202

40	Benzyl acetate
100	Terpineol
20	Bromstyrole
60	Phenyl ethyl alcohol
100	Amyl salicylate
100	Geraniol, Java
400	Clove
50	Cassia
70	Iso-eugenol
30	Musk xylene
30	Styrax resin
1000	

CHYPRE

Chypre Soap, no. 1203

400	Bergamot
100	Methyl ionone
100	Geranium, Bourbon
5	Civet absolute
50	Orris resin
5	Castoreum absolute
50	Coumarin
30	Musk xylene
50	Oakmoss
30	Patchouli
80	Santal
100	Peru balsam
1000	

Chypre Soap, no. 1204

50	Benzyl acetate
100	Rosewood
150	Cedarwood
100	Linalyl acetate
80	Ionone
20	Ethyl cinnamate
50	Amyl cinnamic aldehyde
50	Iso-eugenol
100	Coumarin
50	Musk xylene
50	Oakmoss
100	Styrax resin

50	Tolu balsam
50	Vetivert
1000	

Odour classification

Top notes	Middle notes	Basic notes
1. Benzyl acetate	16. Eugenol	79. Civet absolute
2. Rosewood	19. Verbena	90. Orris resin
3. Sassafras	21. Ionone	100. Amyl cinnamic aldehyde
4. Lavender	22. Clove	Benzoin resin
Phenyl ethyl alcohol	24. Ethyl cinnamate	Castoreum absolute
	Geranium, African	Coumarin
6. Bergamot	45. Cassia	Cypress
8. Amyl salicylate		Iso-eugenol
Cedarwood		Labdanum resin
Lemon		Musk xylene
9. Spike lavender		Oakmoss
10. Linalyl acetate		Olibanum resin
11. Sweet orange		Patchouli
14. Cananga		Peru balsam
Methyl ionone		Santal
		Styrax resin
		Tolu balsam
		Vetivert

Curd, no. 1205

300	Citronella, Ceylon
50	Cassia
150	Lemon
50	Caraway
250	Geranium, Bourbon
100	Bergamot
30	Peppermint
70	Terpineol
1000	

Curd, no. 1206

200	Safrole
300	Geraniol
500	Terpineol
1000	

COLOGNE

Cologne Soap, no. 1207

200	Petitgrain para
50	Lavender
200	Bergamot
150	Lemon
50	Lemongrass
40	Methyl anthranilate
60	Rosemary
100	Geranium, Bourbon
100	Citral
50	Methyl naphthal ketone
1000	

Odour classification

Top notes	Middle notes	Basic notes
1. Benzyl acetate	21. Methyl anthranilate	77. Methyl naphthyl ketone
2. Rosewood	Rosemary	100. Benzoin resin
3. Petitgrain para	22. Clove	Coumarin
Terpineol	24. Geranium	Musk xylene
4. Citronellol	Geranyl acetate	Styrax resin
Lavender	30. Gingergrass	
Phenyl ethyl alcohol	60. Citral	
5. Terpinyl acetate		
6. Bergamot		
7. Geraniol, Java		
8. Citronella, Ceylon		
Lemon		
9. Spike lavender		
Thyme, red		
10. Linalyl acetate		
11. Sweet orange		
12. Methyl cinnamate		
14. Lemongrass		

Cologne Soap, no. 1208

200	Petitgrain para
100	Benzyl acetate
200	Terpinyl acetate
100	Geraniol, Java
100	Citronella, Ceylon

50	Spike lavender
50	Methyl anthranilate
100	Citral
70	Methyl naphthyl ketone
30	Musk xylene
1000	

CYCLAMEN

Odour classification

Top notes	Middle notes	Basic notes
1. Benzyl acetate	15. Guaiac wood	79. Civet absolute
2. Rosewood	21. Anisic aldehyde Ionone	80. Hydroxy citronellal
3. Terpineol	60. Citral	89. Cyclamen aldehyde
4. Bromstyrole Citronellol Phenyl ethyl alcohol		90. Orris resin
		91. Undecalactone
6. Bergamot Phenyl methyl carbinyl acetate		100. Amyl cinnamic aldehyde
		Benzoin resin
7. Geraniol, Java		Coumarin
8. Amyl salicylate		Musk xylene
10. Linalyl acetate		Oakmoss
12. Methyl heptine carbonate		Olibanum resin
		Phenyl acetic acid
14. Cananga		Santal
		Styrax resin
		Vetivert

Cyclamen Soap, no. 1209

100	Rosewood
100	Terpineol
100	Citronellol
50	Bergamot
50	Cananga
50	Guaiac wood
150	Ionone
200	Hydroxy citronellal
20	Cyclamen aldehyde
30	Amyl cinnamic aldehyde

Cyclamen Soap, no. 1209 (*continued*)

10	Oakmoss
40	Musk xylene
100	Styrax resin
1000	

Cyclamen Soap, no. 1210

100	Rosewood
100	Terpineol
200	Geraniol, Java
50	Cananga
50	Anisic aldehyde
100	Ionone
200	Hydroxy citronellal
20	Cyclamen aldehyde
30	Amyl cinnamic aldehyde
20	Phenyl acetic acid
30	Musk xylene
100	Styrax resin
1000	

FOUGÈRE

Fern Soap, no. 1211

10	Methyl benzoate
30	Rosewood
100	Lavender
200	Bergamot
50	Amyl salicylate
100	Lavendin
150	Linalyl acetate
20	Cananga
30	Rosemary
100	Coumarin
50	Musk xylene
50	Oakmoss
10	Patchouli
100	Santal
1000	

Fern Soap, no. 1212

20	Methyl salicylate
30	Sassafras
350	Terpinyl acetate
150	Spike lavender
60	Cananga
50	Anisic aldehyde
50	Dimethyl hydroquinone
50	Coumarin
50	Musk xylene
40	Oakmoss
100	Styrax resin
50	Vetivert
1000	

Odour classification

Top notes	Middle notes	Basic notes
1. Benzyl acetate	15. Acet anisol	100. Amyl cinnamic aldehyde
Methyl benzoate	21. Anisic aldehyde	Benzophenone
2. Methyl salicylate	Ionone	Coumarin
Rosewood	Myrrh resin	Musk xylene
3. Methyl acetophenone	Rosemary	Oakmoss
Sassafras	24. Geranium	Patchouli
4. Citronellol	27. Dimethyl hydroquinone	Peru balsam
Eucalyptus	45. Bornyl acetate	Santal
Lavender		Styrax resin
5. Terpinyl acetate		Tolu balsam
6. Bergamot		Vanillin
7. Geraniol, Java		Vetivert
8. Amyl salicylate		
Cedarwood		
Lemon		
9. Spike lavender		
Thyme, red		
10. Lavandin		
Linalyl acetate		
12. Petitgrain, French		
14. Cananga		

Glycerine and Cucumber, no. 1213

300	Bergamot
100	Linalol
250	Petitgrain para

Glycerine and Cucumber, no. 1213 (*continued*)

100	Palmarosa
100	Clove
30	Benzyl acetate
20	Geranium, Bourbon
30	Benzoin resin
30	Peru balsam
10	Musk xylene
20	Coumarin
10	Patchouli
1000	

Glycerine and Cucumber, no. 1214

220	Ceylon citronella
200	Terpineol
50	Lemongrass
250	Rosewood
100	Benzyl acetate
100	Clove
30	Styrax resin
50	Musk xylene
1000	

HELIOTROPE

Heliotrope Soap, no. 1215

10	Benzaldehyde
40	Benzyl acetate
10	Bromstyrole
90	Phenyl ethyl alcohol
100	Geraniol, Java
50	Sweet orange
300	Heliotropin
50	Anisic aldehyde
50	Ionone
100	Benzoin resin
50	Musk xylene
100	Peru balsam
50	Vanillin
1000	

Odour classification

Top notes	Middle notes	Basic notes
1. Benzaldehyde	15. Heliotropin	77. Methyl naphthyl ketone
Benzyl acetate	21. Anisic aldehyde	
4. Bromstyrole	Ionone	90. Orris resin
Phenyl ethyl alcohol	22. Clove	91. Undecalactone
	24. Ethyl cinnamate	100. Amyl cinnamic aldehyde
6. Bergamot	29. Geranium, Bourbon	Benzoin resin
7. Geraniol, Java		Coumarin
8. Amyl salicylate		Musk xylene
11. Sweet orange		Patchouli
14. Cananga		Peru balsam
		Santal
		Styrax resin
		Vanillin
		Vetivert

Heliotrope Soap, no. 1216

10	Benzaldehyde
70	Benzyl acetate
20	Bromstyrole
150	Geraniol, Java
50	Cananga
300	Heliotropin
50	Anisic aldehyde
100	Ionone
40	Clove
10	Ethyl cinnamate
50	Musk xylene
100	Styrax resin
50	Vanillin
1000	

Herb, no. 1217

100	Thyme, red
100	Rosemary
200	Lavender, spike
50	Fennel
50	Spearmint
300	Geranium, Bourbon
50	Cassia

Herb, no. 1217 (*continued*)

100	Clove
50	Camomile
1000	

Herb, no. 1218

500	Herb compound, no. 1217
100	Citronella, Java
300	Terpineol
100	Rosewood
1000	

Honey, no. 1219

200	Bergamot
200	Rosewood
100	Geranium, Algerian
100	Miel compound, Vol. I
50	Amyl cinnamic aldehyde
10	Civet absolute
60	Peru balsam
30	Musk xylene
150	Phenylethyl alcohol
100	Terpineol
1000	

Honey, no. 1220

450	Citronella, Ceylon
200	Lemon
100	Cinnamon leaf
50	Bay
20	Iso-butyl phenylacetate
100	Phenylethyl alcohol
30	Phenylacetic acid
50	Musk residues
1000	

HYACINTH

Hyacinth Soap, no. 1221

200	Benzyl acetate
200	Rosewood
50	Bromstyrole
100	Phenyl ethyl alcohol
30	Methyl cinnamate
50	Cananga
50	Ionone
70	Clove
50	Geranium, Bourbon
50	Galbanum resin
50	Amyl cinnamic aldehyde
100	Styrax resin
1000	

Odour classification

Top notes	Middle notes	Basic notes
1. Benzaldehyde Benzyl acetate	15. Cinnamyl acetate Styrax oil	85. Phenyl acetaldehyde dimethyl acetal
2. Phenyl ethyl acetate Rosewood	21. Ionone Methyl anthranilate 22. Clove	89. Cyclamen aldehyde
3. Terpineol	24. Ethyl cinnamate Geranium	90. Galbanum resin Opoponax resin
4. Bromstyrole Citronellol Phenyl ethyl alcohol		100. Amyl cinnamic aldehyde Iso-eugenol
6. Bergamot Phenyl methyl carbinyl acetate		Musk xylene Olibanum resin Hydro cinnamic aldehyde
7. Methyl octine carbonate		Styrax resin Vetivert
8. Geraniol palmarosa Iso-butyl salicylate		
10. Linalyl acetate		
11. Opoponax oil Sweet orange		
12. Methyl cinnamate Methyl heptine carbonate		
14. Cananga		

Hyacinth Soap, no. 1222

100	Benzyl acetate
300	Rosewood
30	Bromstyrole
100	Phenyl ethyl alcohol
200	Geraniol, Java
50	Cananga
50	Clove
20	Galbanum resin
50	Amyl cinnamic aldehyde
30	Musk xylene
70	Styrax resin
1000	

JASMIN

Odour classification

Top notes	Middle notes	Basic notes
1. Benzyl acetate	15. Cinnamyl acetate	77. Methyl naphthyl ketone
2. Phenyl ethyl acetate	21. Indole	79. Civet absolute
Rosewood	Ionone	80. Hydroxy citronellal
3. Terpineol	Methyl anthranilate	88. Ethyl methyl phenyl glycidate
Petitgrain para	22. Clove	89. Cyclamen aldehyde
4. Phenyl ethyl alcohol	24. Geranium	90. Orris resin
6. Bergamot	Geranyl acetate	91. Undecalactone
7. Geraniol, Java		100. Amyl cinnamic aldehyde
8. Amyl salicylate		Benzoin resin
10. Linalyl acetate		Coumarin
11. Sweet orange		Iso-eugenol
12. Methyl cinnamate		Labdanum resin
13. Paracresyl phenylacetate		Liquidambar resin
14. Cananga		Musk xylene
Methyl ionone		Phenyl acetic acid
		Santal
		Styrax resin
		Tolu balsam
		Vetivert

Jasmin Soap, no. 1223

200	Benzyl acetate
100	Rosewood
80	Petitgrain para
100	Phenyl ethyl alcohol
100	Bergamot
70	Amyl salicylate
25	Sweet orange
100	Cananga
50	Clove
3	Civet absolute
50	Orris resin
2	Undecalactone
40	Amyl cinnamic aldehyde
50	Liquidambar resin
30	Musk xylene
1000	

Jasmin Soap, no. 1224

200	Benzyl acetate
200	Rosewood
100	Terpineol
150	Petitgrain para
50	Amyl salicylate
50	Cananga
50	Ionone
50	Methyl anthranilate
20	Clove
50	Amyl cinnamic aldehyde
30	Musk xylene
50	Styrax resin
1000	

LAVENDER

Lavender Soap, no. 1225

70	Bergamot
50	Geraniol, Java
20	White thyme
100	Spike lavender
130	Lavender
100	Rosemary
10	Clove

Lavender Soap, no. 1225 (*continued*)

30	Orris resin
70	Benzoin resin
50	Coumarin
30	Musk xylene
30	Oakmoss
10	Patchouli
30	Santal
1000	

Odour classification

Top notes	Middle notes	Basic notes	
1. Benzyl acetate	21. Methyl anthranilate	90.	Orris resin
1. Rosewood	Rosemary	100.	Benzoin resin
3. Petitgrain para	22. Clove		Coumarin
Terpineol	24. Benzylidene		Labdanum resin
4. Lavender	acetone		Musk xylene
5. Terpinyl acetate	Geranyl acetate		Oakmoss
6. Bergamot	29. Geranium, Bourbon		Patchouli
7. Geraniol, Java	45. Bornyl acetate		Peru balsam
Thyme, white			Santal
9. Peppermint			Styrax resin
Spike lavender			Vetivert
10. Lavendin			
Linalyl acetate			
14. Cananga			

Lavender Soap, no. 1226

50	Benzyl acetate
70	Rosewood
30	Petitgrain para
100	Terpinyl acetate
200	Spike lavender
300	Lavendin
50	Rosemary
100	Coumarin
40	Musk xylene
10	Oakmoss
50	Vetivert
1000	

LILAC

Odour classification

Top notes	Middle notes	Basic notes
1. Benzaldehyde 　Benzyl acetate 2. Linalol 　Rosewood 3. Methyl 　　acetophenone 　Paracresyl methyl 　　ether 　Petitgrain para 　Terpineol 4. Bromstyrole 　Citronellol 　Phenyl ethyl 　　alcohol 6. Bergamot 7. Geraniol, Java 8. Amyl salicylate 10. Linalyl acetate 12. Methyl cinnamate 14. Cananga	15. Acet anisol 　　Heliotropin 21. Anisic aldehyde 　　Indole 　　Ionone 　　Methyl anthranilate 22. Clove 　　Phenyl cresyl oxide 24. Geranium	65. Cinnamic alcohol 77. Methyl naphthyl 　　ketone 80. Hydroxy 　　citronellal 85. Phenyl 　　acetaldehyde 　　dimethyl acetal 89. Cyclamen 　　aldehyde 90. Orris resin 91. Undecalactone 100. Amyl cinnamic 　　aldehyde 　Benzoin resin 　Coumarin 　Hydro cinnamic 　　aldehyde 　Iso-eugenol 　Musk xylene 　Olibanum resin 　Peru balsam 　Phenyl acetic acid 　Santal 　Styrax resin 　Tolu balsam

Lilac Soap, no. 1227

50	Benzyl acetate
20	Rosewood
100	Terpineol
10	Bromstyrole
300	Phenyl ethyl alcohol
20	Geraniol, Java
10	Methyl cinnamate
20	Cananga
50	Heliotropin
30	Anisic aldehyde
10	Methyl anthranilate
100	Hydroxy citronellal

Lilac Soap, no. 1227 (*continued*)

10	Cyclamen aldehyde
30	Amyl cinnamic aldehyde
10	Iso-eugenol
30	Musk xylene
100	Peru balsam
100	Styrax resin
1000	

Lilac Soap, no. 1228

30	Benzyl acetate
50	Petitgrain para
200	Terpineol
350	Phenyl ethyl alcohol
50	Cananga
50	Anisic aldehyde
20	Methyl anthranilate
100	Hydroxy citronellal
10	Phenyl acetaldehyde dimethyl acetal
10	Cyclamen aldehyde
50	Amyl cinnamic aldehyde
30	Musk xylene
50	Tolu balsam
1000	

LILY

Lily Soap, no. 1229

50	Benzyl acetate
200	Rosewood
200	Terpineol
200	Citronellol
100	Bergamot
50	Cananga
50	Heliotropin
30	Ionone
100	Hydroxy citronellal
10	Cyclamen aldehyde
20	Amyl cinnamic aldehyde
10	Coumarin
80	Styrax resin
1000	

Odour classification

Top notes	Middle notes	Basic notes

1. Benzaldehyde
   Benzyl acetate
2. Rosewood
3. Terpineol
4. Bromstyrole
   Citronellol
   Phenyl ethyl
         alcohol
6. Bergamot
7. Geraniol, Java
8. Citronella Ceylon
10. Linalyl acetate
11. Sweet orange
12. Methyl cinnamate
    Methyl heptine
         carbonate
    Petitgrain, French
    Phenyl ethyl
         cinnamate
13. Paracresyl
         phenylacetate
14. Cananga
	15. Cinnamyl acetate
	    Heliotropin
	21. Anisic aldehyde
	    Ionone
	    Methyl anthranilate
	29. Geranium, Bourbon
		65. Cinnamic alcohol
		77. Methyl naphthyl
		         ketone
		79. Civet absolute
		80. Hydroxy
		         citronellal
		89. Cyclamen
		         aldehyde
		91. Undecalactone
		100. Amyl cinnamic
		         aldehyde
		Benzoin resin
		Coumarin
		Iso-eugenol
		Santal
		Styrax resin
		Vetivert

Lily Soap, no. 1230

40	Benzyl acetate
100	Rosewood
100	Terpineol
200	Citronellol
100	Cananga
100	Heliotropin
50	Ionone
200	Hydroxy citronellal
10	Cyclamen aldehyde
50	Amyl cinnamic aldehyde
50	Styrax resin
1000	

May Blossom, no. 1231

200	Anisic aldehyde
50	Methyl acetophenone

May Blossom, no. 1231 (*continued*)

100	Bergamot
300	Rosewood
50	Petitgrain
30	Bromstyrole
50	Geranium, Bourbon
40	Amyl cinnamic aldehyde
20	Styrax resin
30	Benzoin resin
10	Civet synthetic
30	Musk xylene
20	Coumarin
70	Heliotropin
1000	

May Blossom, no. 1232

150	Anisic aldehyde
50	Iso-butyl phenylacetate
200	Rosewood
300	Ceylon citronella
20	Styrax resin
40	Heliotropin
40	Musk xylene
20	Bromstyrole
180	Terpineol
1000	

Musk, no. 1233

20	Musk ambrette
30	Musk ketone
40	Vetivert
100	Santal
200	Methyl ionone
70	Liquidambar resin
100	Bergamot
200	Geranium, Bourbon
100	Musk extract, 3 per cent
5	Civet absolute
5	Castoreum absolute
100	Terpineol
30	Cassia oil
1000	

Musk, no. 1234

100	Geranium, Bourbon
100	Clove
30	Cassia
60	Labdanum resin
10	Castoreum absolute
50	Sandal
150	Cedarwood
30	Ionone residues
20	Vetivert
150	Musk xylene
150	Benzyl benzoate
150	Terpineol
1000	

NARCISSUS

Odour classification

Top notes	Middle notes	Basic notes
1. Benzyl acetate	15. Acet anisol	77. Methyl naphthyl ketone
2. Phenyl ethyl acetate	21. Indole Ionone	100. Amyl cinnamic aldehyde
Rosewood	Methyl anthranilate	Benzoin resin
3. Paracresyl acetate	22. Clove	Coumarin
Paracresyl methyl ether	24. Geranium Geranyl acetate	Iso-eugenol
Petitgrain para		Labdanum resin
Terpineol		Musk xylene
4. Bromstyrole		Peru balsam
Citronellol		Styrax resin
Phenyl ethyl alcohol		
6. Bergamot		
7. Geraniol, Java		
10. Linalyl acetate		
12. Methyl cinnamate		
13. Paracresyl phenylacetate		
14. Cananga		

Narcissus Soap, no. 1235

200	Benzyl acetate
200	Rosewood
70	Paracresyl acetate
100	Terpineol
10	Bromstyrole
40	Phenyl ethyl alcohol
100	Geraniol, Java
100	Cananga
1	Indole
20	Methyl anthranilate
40	Clove
9	Amyl cinnamic aldehyde
30	Musk xylene
80	Styrax resin
1000	

Narcissus Soap, no. 1236

200	Benzyl acetate
150	Rosewood
20	Paracresyl acetate
200	Terpineol
100	Phenyl ethyl alcohol
150	Geraniol, Java
50	Cananga
50	Acet anisol
1	Indole
20	Methyl naphthyl ketone
9	Iso-eugenol
50	Peru balsam
1000	

OPOPONAX

Opoponax Soap, no. 1237

100	Benzyl acetate
100	Petitgrain para
40	Phenyl ethyl alcohol
150	Bergamot
50	Cananga
50	Myrrh resin
100	Geranium, Bourbon

50	Opoponax resin
200	Coumarin
10	Castoreum absolute
20	Musk xylene
30	Patchouli
50	Santal
50	Vanillin
1000	

Odour classification

Top notes	Middle notes	Basic notes
1. Benzyl acetate	21. Ionone	90. Opoponax resin
2. Phenyl ethyl acetate	Methyl anthranilate	Orris resin
Rosewood	Myrrh	100. Amyl cinnamic aldehyde
3. Petitgrain para Terpineol	22. Clove	Benzoin resin
4. Citronellol	29. Geranium, Bourbon	Castoreum absolute
Phenyl ethyl alcohol	60. Citral	Coumarin
		Musk xylene
6. Bergamot		Olibanum resin
7. Geraniol, Java		Patchouli
8. Amyl salicylate		Peru balsam
Citronella, Ceylon		Santal
11. Opoponax oil		Styrax resin
14. Cananga		Tolu balsam
		Vanillin

Opoponax Soap, no. 1238

100	Benzyl acetate
100	Petitgrain para
50	Phenyl ethyl alcohol
100	Terpinyl acetate
100	Geraniol, Java
50	Cananga
50	Opoponax resin
100	Coumarin
30	Musk xylene
20	Patchouli
100	Peru balsam
100	Styrax resin
100	Tolu balsam
1000	

Palm and Olive Oils, no. 1239

50	Bromstyrole
250	Rosewood
30	Cassia
70	Clove
100	Lavender
50	Patchouli
150	Cedarwood
100	Petitgrain
150	Geranium, Bourbon
50	Musk xylene
1000	

Palm and Olive Oils, no. 1240

500	Palm compound, no. 1239
200	Terpineol
300	Lavender, spike
1000	

Pine Bouquet, no. 1241

200	Bornyl acetate
100	Geranium, Bourbon
40	Cassia
50	Clove
50	Petitgrain
250	Lavender
200	Bergamot
50	Terebene
20	Musk xylene
30	Coumarin
10	Patchouli
1000	

Pine Bouquet, no. 1242

50	Borneol
100	Bornyl acetate
100	Cedarwood
200	Lavender, spike
50	Methyl anthranilate
100	Rosewood
200	Terpineol
100	Citronella, Java
50	Terebene

30	Musk residues
20	Coumarin
1000	

ROSE

Rose Soap, no. 1243

200	Citronellol
200	Phenyl ethyl alcohol
200	Geraniol palmarosa
20	Verbena
50	Ionone
10	Clove
150	Geranium, Bourbon
50	Orris resin
30	Musk xylene
50	Santal
30	Trichlor phenyl methyl carbinyl acetate
10	Vetivert
1000	

Odour classification

Top notes	Middle notes	Basic notes
1. Benzyl acetate	15. Citronella, Java	79. Civet absolute
2. Rosewood	Guaiac wood	85. Phenyl acetaldehyde dimethyl acetal
3. Benzyl cinnamate Petitgrain para Terpineol	19. Verbena 21. Ionone 22. Clove	90. Orris resin
4. Citronellol Phenyl ethyl alcohol	24. Ethyl cinnamate Geranium Geranyl acetate	91. Undecalactone 100. Amyl cinnamic aldehyde
6. Bergamot	30. Gingergrass	Benzoin resin
7. Geraniol, Java	45. Cassia	Benzophenone
8. Geraniol palmarosa Lemon	60. Citral	Coumarin Labdanum resin
10. Diphenyl methane Diphenyl oxide		Musk xylene Patchouli
12. Methyl cinnamate		Phenyl acetic acid
14. Cananga Methyl ionone Palmarosa		Santal Styrax resin Trichlor phenyl methyl carbinyl acetate Vetivert

Rose Soap, no. 1244

30	Benzyl acetate
300	Phenyl ethyl alcohol
30	Diphenyl oxide
350	Palmarosa
100	Citronella, Java
10	Ethyl cinnamate
10	Cassia
50	Gingergrass
10	Benzophenone
30	Musk xylene
10	Patchouli
20	Phenyl acetic acid
50	Styrax resin
1000	

SANTAL

Odour classification

Top notes	Middle notes	Basic notes
2. Rosewood	22. Clove	79. Civet absolute
3. Terpineol	24. Geranium	90. Orris resin
6. Bergamot	41. Mace	100. Coumarin
7. Geraniol, Java	45. Cassia	Musk xylene
8. Cedarwood		Oakmoss
10. Lavendin		Patchouli
11. Nutmeg		Peru balsam
14. Cananga		Santal
		Styrax resin
		Vetivert

Santal Soap, no. 1245

100	Rosewood
100	Geraniol, Java
30	Clove
50	Geranium, Bourbon
50	Orris resin
70	Coumarin
30	Musk xylene
20	Patchouli

100	Peru balsam
400	Santal
50	Vetivert
1000	

Santal Soap, no. 1246

100	Terpineol
200	Cedarwood
50	Cassia
100	Coumarin
50	Musk xylene
300	Santal
100	Vetivert
100	Styrax resin
1000	

Shaving Stick, no. 1247

300	Geranium, Algerian
170	Rosewood
20	Almond, bitter S.A.P.
20	Vetivert
100	Terpineol
70	Phenyl ethyl alcohol
100	Benzyl acetate
100	Santal
50	Benzoin resin
50	Peru balsam
20	Styrax
1000	

Shaving Stick, no. 1248

250	Citronella, Ceylon
300	Lavender, spike
100	Santal
60	Coumarin
40	Amyl salicylate
50	Musk residues
200	Terpineol
1000	

SWEET PEA

Sweet Pea Soap, no. 1249

50	Benzyl acetate
250	Rosewood
150	Petitgrain para
150	Terpineol
30	Bromstyrole
20	Iso-butyl phenylacetate
10	Methyl heptine carbonate
50	Cananga
10	Anisic aldehyde
50	Methyl anthranilate
50	Methyl naphthyl ketone
50	Phenyl acetaldehyde dimethyl acetal
30	Amyl cinnamic aldehyde
30	Musk xylene
20	Phenyl acetic acid
50	Styrax resin
1000	

Odour classification

Top notes	Middle notes	Basic notes
1. Benzyl acetate	15. Guaiac wood	77. Methyl naphthyl ketone
2. Phenyl ethyl acetate	21. Anisic aldehyde	85. Phenyl acetaldehyde dimethyl acetal
Rosewood	Ionone	
3. Methyl acetophenone	Methyl anthranilate	91. Undecalactone
Petitgrain para	22. Clove	100. Amyl cinnamic aldehyde
Terpineol	24. Benzylidene acetone	Benzoin resin
4. Bromstyrole	Geranium	Musk xylene
Citronellol		Phenyl acetic acid
Phenyl ethyl alcohol		Santal
6. Bergamot		Styrax resin
Iso-butyl phenylacetate		Tolu balsam
7. Geraniol, Java		Vetivert
8. Citronella, Ceylon		
10. Diphenyl oxide		
11. Sweet orange		
12. Methyl heptine carbonate		
14. Cananga		

Sweet Pea Soap, no. 1250

100	Benzyl acetate
200	Rosewood
200	Petitgrain para
100	Terpineol
50	Bromstyrole
50	Iso-butyl phenylacetate
4	Methyl heptine carbonate
50	Cananga
20	Benzylidene acetone
50	Methyl naphthyl ketone
1	Undecalactone
50	Amyl cinnamic aldehyde
25	Musk xylene
50	Styrax resin
50	Tolu balsam
1000	

Transparent, no. 1251

450	Geranium, Bourbon
40	Patchouli
10	Vetivert
100	Clove
30	Cassia
20	Caraway
100	Lavender
60	Orris resin
40	Methyl ionone
20	Coumarin
30	Musk xylene
100	Terpineol
1000	

Transparent, no. 1252

100	Citronella, Java
50	Cassia
50	Clove
100	Thyme, red
50	Sassafras
250	Lavender, spike
200	Geraniol
180	Cedarwood
20	Musk xylene
1000	

Trèfle, no. 1253

100	Geranium, Bourbon
10	Oakmoss
100	Rosewood
50	Peru balsam
400	Amyl salicylate
60	Benzyl acetate
30	Amyl cinnamic aldehyde
100	Bergamot
50	Phenylethyl alcohol
100	Cananga
1000	

Trèfle, no. 1254

400	Amyl salicylate
100	Geraniol, Java
50	Phenylethyl alcohol
20	Bromstyrole
250	Terpineol
50	Benzyl acetate
30	Musk residues
100	Rosewood
1000	

VERBENA

Odour classification

Top notes	Middle notes	Basic notes
1. Benzyl acetate	15. Citronella, Java	90. Olibanum resin
Methyl benzoate	19. Verbena	100. Benzoin resin
2. Rosewood	21. Ionone	Coumarin
3. Petitgrain para	Methyl anthranilate	Iso-eugenol
Terpineol	22. Clove	Musk xylene
4. Bromstyrole	24. Geranium	Oakmoss
Phenyl ethyl alcohol	Geranyl acetate	Patchouli
	30. Gingergrass	Santal
6. Bergamot	45. Cassia	Tolu balsam
7. Geraniol, Java	60. Citral	Vetivert
8. Lemon		
9. Spike lavender		
10. Lavendin		
Linalyl acetate		
14. Cananga		
Lemongrass		
Palmarosa		

Verbena Soap, no. 1255

30	Benzyl acetate
5	Methyl benzoate
100	Petitgrain para
100	Bergamot
100	Geraniol, Java
50	Lemon
200	Lemongrass
100	Ionone
50	Clove
50	Geranium, Bourbon
80	Citral
40	Olibanum resin
50	Benzoin resin
30	Musk xylene
10	Oakmoss
5	Patchouli
1000	

Verbena Soap, no. 1256

100	Benzyl acetate
400	Lemongrass
200	Citronella, Java
50	Ionone
30	Methyl anthranilate
10	Cassia
100	Citral
20	Olibanum resin
30	Musk xylene
10	Coumarin
20	Iso-eugenol
30	Vetivert
1000	

VIOLET
Odour classification

Top notes	Middle notes	Basic notes
1. Benzyl acetate	15. Guaiac wood	79. Civet absolute
2. Rosewood	21. Anisic aldehyde	90. Orris resin
6. Bergamot	Ionone	100. Benzoin resin
7. Geraniol, Java	22. Benzyl iso-eugenol	Iso-eugenol
8. Cedarwood		Musk xylene
12. Methyl heptine carbonate		Santal
		Styrax resin
14. Cananga		Vetivert
Methyl ionone		
Mimosa absolute		

Violet Soap, no. 1257

100	Rosewood
150	Bergamot
10	Methyl heptine carbonate
40	Cananga
300	Methyl ionone
20	Mimosa absolute
30	Anisic aldehyde
200	Ionone
50	Orris resin
10	Iso-eugenol
30	Musk xylene
30	Styrax resin
30	Vetivert
1000	

Violet Soap, no. 1258

100	Benzyl acetate
150	Rosewood
50	Geraniol, Java
200	Cedarwood
50	Cananga
50	Benzyl iso-eugenol
300	Ionone
30	Musk xylene
70	Santal
1000	

The above compounds may be used as much as price will allow. For strength it may be necessary to employ 2 per cent, but about 1 per cent will generally be found ample.

Antiseptic and medicated soaps

Antiseptic and medicated soaps are made up with all kinds of medicaments, such as iodoform, thymol, phenol, betanaphthol, sulphur, etc. Many of the essential oils used in perfuming the soaps are highly antiseptic. In order to judge their antiseptic value, Cavel studied their action on ordinary beef-tea which he had previously infected with water taken from the collecting tank of a sewerage system. He determined for the following volatile oils, the dilution per 1000, at which they no longer showed antiseptic value. It will be noted that phenol was 5·6 and is therefore a long way down the list:

Thyme	0·7	Peppermint	2·5
Organium	1·0	Rose geranium	2·7
Sweet orange	1·2	Vetivert	2·7
Verbena	1·6	Bitter almond	2·8
Cassia	1·7	Eucalyptus	2·25
Rose	1·8	Gaultheria	3·0
Clove	2·0	Palmarosa	3·1
Spike	3·5	Lavender	5·0
Star anise	3·7	Balm	5·2
Orris	3·8	Ylang-ylang	5·6
Ceylon cinnamon	4·0	*Phenol*	5·6
Canadian snake root	4·0	Fennel, sweet	6·4
Birch	4·1	Lemon	7·0
Anise	4·2	Sassafras	7·5
Rosemary	4·3	Limes	8·4
Cumin	4·5	Angelica	10·0
Neroli	4·75	Patchouli	15·0

A highly antiseptic perfume, therefore, may be compounded on the following lines:

No. 1259

50	Thyme oil
100	Cassia oil
100	Clove oil
50	Eucalyptus oil
200	Rose geranium oil
20	Vetivert oil
80	Rosemary oil
200	Lavender oil
200	Lemon oil
1000	

Perfuming boxes—when packing wrapped soap some makers spray the boxes with perfume—an example of which is given:

No. 1260

30	Bergamot oil
40	Geranium oil, Bourbon
30	Rose, synthetic
10	Jasmin, synthetic
20	Tuberose, synthetic
5	Oakmoss

No. 1260 (*continued*)

20	Benzyl cinnamate
20	Coumarin
10	Musk ambrette
100	Tincture of civet, 3 per cent
715	Tincture of benzoin, 10 per cent
1000	

CHAPTER NINE

Tobacco Flavours

The use of tobacco originated in America, and the name is probably derived from the original word used to describe the pipes employed by the Carib Indians for smoking the weed. It was introduced into Western Europe about 1500, first into Spain and subsequently into England by Sir Walter Raleigh in 1589. Despite the opposition against its use by Royalty and religious dignitaries, the habit spread rapidly to the Orient and thence to the East. Today the manufacture of cigarettes and tobacco is one of our important industries, and the revenue obtained through its large consumption is of considerable assistance in helping the Chancellor of the Exchequer to balance his budgets. The tobacco industry may be divided broadly into two stages:

1. The cultivation of the plant, together with drying, curing, prizing, etc.
2. The conversion of the leaf into tobacconists' commodities, involving the perfuming or flavouring of the imported raw material.

The plant

The tobacco plant thrives best in the tropics, and is cultivated particularly in Cuba, Central America, Borneo, Sumatra, Rhodesia, Turkey, Bulgaria, and the Philippines. It can be grown also in temperate countries such as Canada, and might even flourish in this country were it not for the restrictions placed upon its cultivation by the Excise authorities. The only place of note where English cultivation has proved successful is situated at Crookham in Hants where the leaf has been grown since 1910. There are some 10 to 15 acres yielding a crop having an average value of £4000. The plant is a member of the N.O. solanaceae, and the genus *Nicotiana* consists of several species, of which the following are of importance:

N. tabacum, L., is indigenous to many parts of South America. It attains a height of 5 or 6 feet and develops large ovate or lanceolate leaves. The flowers are pale pink, tubular or bell-shaped. This species yields the common Virginian tobacco of commerce. When fresh the odour of the plant is narcotic, its taste is nauseous and bitter. The characteristic tobacco smell is developed in the course of curing, etc.

N. rustica, L., is cultivated in Southern Europe, Turkey, and the Levant. The plant seldom exceeds 2 or 3 feet in height and its leaves are roundish ovate. The flowers are green, and the flavour after curing is much milder than the above. The leaves constitute the raw material for Turkish and Syrian cigarettes. It burns too rapidly for use in pipe tobacco.

N. Persica, Lindley, is grown mainly in Persia. The leaves of this species are oblong spatulate and the flowers white and aromatic. It constitutes the raw material for the tobacco of Shiraz.

N. nepanda, Willd., is cultivated principally in Havana, where it is used in the manufacture of a special variety of cigar.

N. multivalvis, Lindley, and *N. quadrivalvis*, Pursh, are also grown for special purposes.

N. acuminata, Graham, is cultivated in the mountainous districts of Northern Syria. The plant is a small one, seldom exceeding a few inches in height, and contrary to the usual practice of cultivators, is allowed to flower. The whole plant above ground is employed and is known as Latakia. Its peculiar odour and characteristic flavour is said to be due to the use, during curing, of a fuel consisting largely of dried camel dung.

Cultivation

Tobacco, like many other substances, varies according to the soil upon which the plant is cultivated. The best results are obtained from well-drained ground rich in organic constituents. Sheltered positions are chosen to prevent the laceration and twisting of the leaves by the wind. The plants are raised from seed, sown in nursery beds during early spring. Those which have developed sufficiently are transplanted to the open fields about two months after sowing—wet weather being preferred. They are set out in rows and the ground requires frequent hoeing to keep down weeds, etc., and allow the plant to grow healthily. After seven or eight weeks the flower buds appear and are at once removed to ensure a finer development of leaf structure. Leaves which are near the ground as well as those at the top of the stem are removed— the number left (from 10 to 20) being judged by the grower to yield a finer crop. Ten to twelve weeks after planting the organism ripens, which is evidenced by the development of a yellowish-green colour. A resinous substance exudes from the leaves, which sometimes become mottled with yellowish spots. At this period the ripest plants are harvested, and cloudy weather is, if possible, chosen for the purpose. The leaves are sometimes removed from the stem but the commoner practice is to cut the stem near the ground and to place it on one side in the sun to wilt. The best quality leaves, known as 'first brights', are taken to the barn on the stem which is split down the centre to within a few inches of the ground. They are thus 'straddled', which preserves the leaves in the best condition.

Curing

Tobacco, like vanilla, owes its characteristic odour and flavour to the curing process, which probably results in the action of some enzyme on a glucoside present in the plant. Two curing processes are employed, and the choice depends upon whether the leaves are pale or dark. If the former, they are 'sun-cured'. If the latter, they are cured by fermentation.

The sun-curing process is effected by one of two methods. The leaves are either (1) exposed to the sun at a temperature of about 22°C for four or five days and afterwards hung in well-ventilated barns until soft and ready for packing, or (2) placed in buildings artificially heated where the temperature is gradually raised from 33°C to 77°C. The control of the temperature requires very skilled judgment, the object being to turn the leaves a good yellow and then fix the colour. When the process is completed the leaves are damped and stripped from the stem. In wet weather the atmosphere is sufficiently moist to dispense with damping. The leaves are subsequently graded and packed.

The fermenting or sweating process is employed for curing most of the tobacco exported to Europe. The stems are hung in barns and the temperature raised quickly to about 80°C. This is maintained for four or five days until the leaves are quite dry and brittle. The doors of the barn are opened on a damp day when the leaves absorb moisture and become quite pliable again. They are then stripped from the stem and sorted, being subsequently bundled into twenties or thereabouts and heaped together on the floor. The temperature within the heap gradually rises until it reaches 55°C when the heap is pulled to pieces to prevent overheating and the heap reformed. The tobacco assumes a uniformly brown tint in about one month, when the process is completed.

The bundles of cured tobacco leaf are now so arranged that they can be pressed by machinery into hogsheads of about 1000 lb. Too great pressure causes blackening of the leaf and must be avoided. The object of prizing in standard weights is to facilitate calculations and examination in bulk. Tobacco is sometimes 'improved' at this stage by the addition of molasses, rum, vanilla, etc.

Constituents

The active principle of tobacco is a liquid volatile alkaloid named *Nicotine* and discovered in 1828 by Posselt and Reimann. Albumin, gums, and resins are also present. An essential oil is present in cured tobacco. Schimmel & Co. distilled 15 kg and obtained a yield of 6 grams of a thick balsamic dark brown oil having an odour of camomile. When tobacco is smoked this volatile oil doubtless undergoes a certain amount of decomposition. Thoms obtained from 20 kg a yield of 75 grams of an ethereal oil which contained a phenol and probably furfurol. Iso-valeric acid and iso-butyl acetic acid were also identified as constituents.

Manufacture

Since the sale of tobacco yields such a handsome return to the State in the form of duty, it naturally follows that the Excise authorities keep a watchful eye on the production of the finished article. Two kinds of factories are in operation:

1. Excise factories, in which duty-paid tobacco only is manufactured under excise supervision. The higher grade products are made here.

2. Bonded factories, in which the finished tobacco is prepared and taxed when issued for sale. These premises are strictly controlled by customs' officials, and the use of heavy sweetening materials is permissible only in them.

The use of flavours or perfumes is allowed in either kind of factory, and it is on the composition of these that a manufacturer often relies for the popularity of his brands. When preparing a new flavour it is always desirable to consult the Excise officer and obtain his permission, especially in view of the stringency required by the law. In order to save the time of the tobacco chemist, it seems desirable that a note should be made of those substances prohibited by the Tobacco Act of 1842. These are: sugar, treacle, molasses, honey-combings or roots of malt, ground or unground rooted grain, unground chicory, lime, sand (not tobacco sand), amber, ochre or other earth, seaweed, ground or powdered wood, moss or weeds, or any leaves or any herbs or plants (not being tobacco leaves or plants). Antiseptics other than acetic acid, added colouring matter, any substance imitating tobacco as an internal wrapper.

The following substances are allowed in the manufacture of tobacco:

1. Acetic acid.
2. Essential oils dissolved in spirit for flavouring *cut* tobacco.
3. Tonka beans up to 3 per cent and orris-root up to 2 per cent for scenting snuff. (This applies to manufacturers only and not to dealers and retailers.)

In amplification of this it will be as well to state that the author has found the Board of Customs and Excise to offer no reasonable objection to any aromatic substance which is readily volatile and does not appreciably increase the weight of the finished product.

Thus alcoholic tinctures of vanilla, orris and tonquin bean have been approved, whereas resinous tinctures such as benzoin, styrax, tolu and Peru balsam have been objected to. Synthetics such as coumarin, vanillin, heliotropin, iso-butyl valerianate, the rose alcohols and the majority of the essential oils have been freely approved, but from conversations with the Government chemist, it would appear that high-boiling synthetics such as benzyl benzoate and ethyl phthalate might be barred. Liquorice can of course be used only in Bond. The Tobacco Act of 1900 restricts the amount of oil (fixed such as olive) to 4 per cent. This is used mainly in roll tobacco.

Cigars

The finest quality leaf for the manufacture of cigars is grown in the Vuelto-Abajo district of Cuba. This is situated close to Havana where the best-known brands are produced. Other grades of tobacco are imported from Jamaica, the Philippines, and the Islands of the Malay Archipelago and these are well suited for medium quality cigars.

The cured leaf is damped to make it pliable. This enables the operator to more easily remove or 'strip' the midrib, which is subsequently used in snuff manufacture. The 'stripped' leaf is now graded according to which part of the cigar it will ultimately form—as for instance—'Wrappers' for the outside finish, 'Bunch wrappers and Fillers' for the inside. The leaf or the inside of the box is now sprayed with perfume and generally all odours are of the cascarilla-coumarin or labdanum types. Weak alcohol is employed as a solvent which is sometimes supplemented with small quantities of brandy or rum. An example is given:

No. 1261

20	Cascarilla oil
30	Cinnamon leaf oil
100	Vanilla extract, 10 per cent
15	Coumarin
5	Rose otto
10	Santalwood oil
70	Brandy
750	Spirit
1000	

This type of flavouring is applicable to Havana cigars, but is supplemented with traces of cinnamon bark oil for Manila cigars. The finished cigar is hand rolled, sorted, labelled, and packed.

Cedarwood boxes were used for this last purpose, but today *Cedrela wood*[1] is generally preferred. This has a pleasant cedar-like odour which is often improved by added perfume sprayed on the box before the cigars are inserted. Such a perfume may be prepared as follows:

No. 1262

5	Cedarwood oil
2	Geranium, Algerian
1	Patchouli oil
2	Santalwood oil
10	Labdanum
1000	Alcohol
1020	

Cigarettes

The cured tobacco is damped to make it pliable, and in the case of Virginia leaf the midrib is removed as above described, while in the Turkish, this is so small as to be unnecessary. The leaf is then passed through a cutting machine adjusted to yield very thin strips. In the case of the Virginian leaf the tobacco is panned or stoved which serves the dual purpose of developing the flavour and removing excessive moisture. This requires expert attention for good results. In the case of Turkish leaf, this part of the process is unnecessary since the delicate flavour would be injured. The tobacco now requires to be perfumed before the hands or machines turn out the finished cigarette. The perfumes are blended on the following lines:

Virginian—First Grade, no. 1263

70	Coumarin
20	Rose otto
10	Clove oil
80	Lavender oil
40	Bergamot oil
30	Cascarilla oil
100	Sweet orange oil
400	Geranium oil
50	Isobutyl valerianate
100	Orris tincture
40	Vanilla absolute
60	Tonka bean absolute
1000	

[1] Consult Vol. I.

Virginian—Second Grade, no. 1264

50	Coumarin
20	Sweet orange oil
30	Bergamot oil
30	Lemon oil
20	Phenylethyl valerianate
5	Cassia oil
50	Geranium oil
15	Clove oil
30	Tonka bean tincture
250	Tincture of vanilla
500	Tincture of orris
1000	

Turkish, no. 1265

5	Rose otto
9	Geranium oil, Bourbon
5	Orange blossom absolute
3	Tuberose compound
2	Jasmin absolute
8	Lavender oil
2	Patchouli oil
1	Methyl phenylacetate
5	Tonka absolute
160	Tincture of vanilla
800	Tincture of orris
1000	

Egyptian, no. 1266

5	Neroli oil
35	Jasmin absolute
10	Cassie absolute
10	Rose absolute
240	Tincture of vanilla
200	Tincture of tonka
500	Spirit
1000	

Tobacco

When the leaf is imported the moisture content seldom exceeds 10 per cent. This is not sufficient, and since the law allows a maximum of 32 per cent, the leaf is damped. This is often carried out by blowing steam into the loosened bundles of tobacco. With the exception of 'Bird's Eye' the midribs of the leaves are removed as soon as they become pliable. The perfume is sprayed on and evenly distributed by turning the leaves over. The leaf is now pressed through a cutting machine which is adjusted for fine, medium, or coarse cut as desired. The drying and developing of the flavour is carried out as indicated above by panning or stoving. The length of time allowed for this is determined by the type of tobacco being manufactured: for instance, shag and Cavendish are allowed to remain on the hot plate longer than flake. Some of the stronger tobaccos are darkened by the addition of olive oil before pressing between hot plates. Pigtail and twist are made from such grades which are eventually spun on special machines. Rum and extract of liquorice are generally added to improve the flavour, since these tobaccos are principally chewed. Plug, cake, and bar tobacco are made by pressure of the leaf in special machines.

The flavours employed for the various types are prepared very much on the basis of coumarin, nutmeg, cascarilla, clove, vanilla, and iso-butyl valerianate. The stronger the tobacco, the higher the proportion used. In the 'chewing' varieties, flavours such as aniseed and caraway are common. A formula illustrative of each type is given:

Flake, no. 1267

50	Coumarin
20	Heliotropin
20	Vanillin
150	Sweet orange oil
20	Cassia oil
30	Clove oil
30	Rose otto
50	Rhodinyl valerianate
50	Cascarilla oil
80	Nutmeg oil
500	Tincture of tonka
1000	

Shag, no. 1268

100	Coumarin
100	Cassia oil
200	Nutmeg oil

Shag, no. 1268 (*continued*)

50	Clove oil
50	Geranium oil, Algerian
50	Cascarilla oil
50	Peru balsam oil
100	Phenylethyl valerianate
100	Tincture of vanilla
200	Tincture of tonka
1000	

Roll (made in Bond), no. 1269

50	Coumarin
30	Anethol
70	Carvone
20	Phenylethyl valerianate
30	Orange oil
600	Rum
200	Liquorice, liquid extract
1000	

Snuff

Two distinct kinds are manufactured.
1. Dry from midribs.
2. Wet or moist from midribs, small leaves and waste.

The former is prepared by damping down the raw materials and placing them aside in bins to ferment. Sometimes alkaline salts are dissolved in the water used for damping but the percentages must be kept within prescribed limits. In certain kinds of dry snuff the fermented mass is subjected to 'toasting' in special furnaces and is subsequently ground.

The latter is prepared in a similar manner, excepting that the temperature during fermentation is now allowed to exceed 55°C and the mass is only completely cured after several weeks. It is subsequently ground.

All snuffs, excepting Welsh and Irish, are perfumed—one well-known brand owes its distinctive odour to a liberal use of bergamot oil. An example follows:

No. 1270

100	Coumarin
100	Lavender oil

400	Bergamot oil
100	Geranium oil
100	Nutmeg oil
50	Cinnamon leaf oil
50	Cascarilla oil
100	Vanilla essence
1000	

CHAPTER TEN

Floral Cachous

The perfuming of the breath by means of mouth pellets is of comparatively recent origin. The ancient peoples knew the flavouring value of several of the plants and their by-products, but utilised them principally as flavours for wine. It is doubtful, however, whether they appreciated their value so much in this direction as they did in the preparations of incense, perfumes, and fumigants.

Mouth pellets in their present form are said to have originated in Italy. There was evidently an attempt to make a cachou by using tragacanth as the base and perfuming it with essential oils and grain musk and these perfumed pellets were very much liked by heavy smokers. An improvement was made in their production by utilising a mixture of liquorice and sugar as the base and spices as well as essential oils as the flavour. It was customary to make them very much as pills and to give them a silver or gilt finish. The following formula will illustrate the type:

No. 1271

10	Peppermint oil
10	Orange oil, sweet
10	Lemon oil
15	Neroli oil
5	Bergamot oil
50	Cinnamon bark, in powder
50	Cloves, in powder
100	Vanilla beans, ground
150	Orris, in powder
200	Powdered sugar

FLORAL CACHOUS

```
 400   Extract of liquorice
 q.s.  Mucilage of acacia
─────
1000
```

Today these have fallen very much out of favour owing to the more attractive flavours and colours of the modern cachou. These are now manufactured either as lozenges or tablets, the former possessing the great advantage of toughness. They therefore retain the flavour to the last thin wafer and do not disintegrate rapidly as tablets are prone to do.

Lozenge-made cachous

As previously indicated by the author, the standards for a really good cachou are as follows:

1. A smooth touch to the tongue.
2. A toughness sufficient to prevent rapid disintegration.
3. A delicate and attractive colour.
4. A persistent flavour approximating to the fragrance of the natural flower.

In order that these characters shall be evident in the finished product, the following points require to be carefully noted: Roughness on the surface of a cachou is generally caused by a too coarse sugar powder. The finest icing obtainable only should be used, and this is produced by repeated grinding and sifting through a fine mesh in a Gardner or other machine. The sugar should all pass through a 160 sifter. The sugar base may be improved, and at the same time slightly cheapened, by the addition of 10 per cent of starch. Toughness is imparted by the use of tragacanth. Acacia alone is very hard and inclined to be brittle. By using a quarter the weight of tragacanth the necessary quantity of acacia is reduced to a minimum. About 4 per cent of acacia and 1 per cent of tragacanth is a reasonable medium. A solution of glucose, 1 in 2, is used to form the mass, and generally 10 to 15 per cent will be found sufficient. The formula for the cachou-base will now read:

No. 1272

```
900  Finest icing sugar (160 mesh)
100  Starch
 40  Acacia, in powder
 10  Tragacanth, in powder
```

No. 1272 (continued)

q.s.	Glucose solution, 50 per cent	
1050		

The colour of the product will naturally be as near as possible to that of the natural flower, for instance: carnation or rose may be coloured pink with a solution of carmoisine, rhodamine, carmine, etc.; jasmin may be coloured yellow with tartrazine; violet or lavender may be tinted a heliotrope shade with methyl violet solution. The best procedure is to make a standard strength solution, say 2 per cent, and, after having found by experiment the quantity of basic colour required, shade or 'top' it with another. For instance, when colouring rose cachous many operators prefer a yellowish-pink tint, which, in their opinion, is a nearer imitation of the colour of a certain type of rose. Carmoisine would be used as the basic colour, and the shading could be conveniently accomplished with tartrazine.

The last character is generally the most important one, and frequently the selling feature of the product. Many of the popular flavours have a basis of vanilla, cinnamon, and cloves, rounded off with rose, patchouli, musk, etc. Synthetic aromatic chemicals and natural isolates, when employed, should be used with much discretion, since they have generally a coarser flavour. They are best employed in small quantities to modify the flavours of other essential oils. An example of a popular type of cachou flavour is appended; it will offer endless possibilities for slight modification to the experimenter:

Bouquet, no. 1273

1	Vanillin
3	Coumarin
1	Benzaldehyde
1	Nutmeg oil
2	Rose otto
2	Cinnamon oil (bark)
2	Cassia oil
8	Clove oil
10	Lavender oil
5	Patchouli oil
20	Musk tincture, 3 per cent
45	Vanilla essence, 10 per cent
100	

Dissolve the solids in the mixed liquids without the direct application of heat, and mature for at least one month before use. Employ $\frac{1}{10}$ per cent to $\frac{1}{4}$ per cent according to taste.

Other flavours having a distinct floral note may be prepared as follows:

Carnation, no. 1274

200	Iso-eugenol
200	Clove oil
50	Ylang-ylang oil
30	Neroli oil
20	Rose otto
300	Benzoin tincture, 10 per cent
150	Musk tincture, 3 per cent
40	Heliotropin
10	Vanillin
1000	

Hawthorn, no. 1275

300	Anisic aldehyde
50	Almond oil, bitter S.A.P.
50	Neroli oil
100	Geranium oil, Bourbon
300	Benzoin tincture, 10 per cent
140	Musk tincture, 3 per cent
10	Orris oil, concrete
30	Coumarin
20	Vanillin
1000	

Heliotrope, no. 1276

200	Heliotropin
20	Vanillin
50	Almond oil, bitter S.A.P.
100	Geranium oil, Bourbon
80	Clove oil
50	Peru balsam oil
300	Vanilla essence, 10 per cent
200	Benzoin tincture, 10 per cent
1000	

Jasmin, no. 1277

20	Jasmin absolute
30	Rose absolute
50	Benzyl acetate
100	Orange oil, sweet
50	Neroli oil
50	Bois de rose oil
700	Vanilla essence, 10 per cent
1000	

Rose, no. 1278

100	Rhodinol
25	Rose otto, Bulgarian
100	Geranium oil, Bourbon
100	Phenylethyl alcohol
150	Clove oil
20	Cassia oil
5	Patchouli oil, English
200	Musk tincture, 3 per cent
100	Benzoin tincture, 10 per cent
200	Vanilla essence, 10 per cent
1000	

Many rose cachous are perfumed with the otto only but they lack body.

Violet, no. 1279

200	Methyl ionone
100	Santalwood oil
100	Clove oil
20	Ylang-ylang oil
30	Bergamot oil
50	Geranium oil, Algerian
20	Orris oil, concrete
30	Heliotropin
450	Vanilla essence, 10 per cent
1000	

Having now obtained all the data to proceed, it only remains to describe briefly the methods of manufacture. Icing sugar easily becomes damp unless very carefully stored. Should it become caked into lumps, coarse and then fine sifting is necessary.

FLORAL CACHOUS

Providing it is purchased of the requisite fineness, the dry, raw material need not be put through a sifting machine. It is generally desirable to put it through a sieve of about 20 mesh when introducing it into the machine or mortar; this ensures the absence of small lumps, which become particularly evident after the base has been coloured. The other solids are passed through the same sieve, the flavour and colour added, and finally the excipient. The mass is allowed to rotate in the machine, or is worked in the mortar, until a plastic substance is obtained. This is then rolled out by hand with a roller and board, or for large operations by machine, sufficient potato starch being employed to prevent the paste adhering to the board. When the correct thickness has been obtained, it is cut up into different shapes and placed on trays to dry (without artificial heating). The cachous are subsequently brushed to remove excess of farina and packed.

Tablet-made cachous

The details already given for perfume and colour also apply to the manufacture of tablets, but the cachou base and process of necessity differ. The basis of the cachou is again icing sugar in fine powder, to which is added varying proportions of lactose. This is employed because it is an excellent absorbent for essential oils, and when large quantities are used for flavouring it prevents the tablets spotting and becoming soft after compression. Powdered acacia is generally used to give the necessary binding, and this is sometimes supplemented by the addition of tragacanth. A generally useful base can be made from these ingredients as follows:

No. 1280

900	Icing sugar
100	Lactose
20	Acacia
10	Tragacanth
q.s.	Syrup
1030	

The two gums may be varied according to whether the finished tablet is required to be harder or tougher. The manufacturing process is carried out by mixing together the powders and adding the necessary quantity of colours in solution. These are placed in a mixing pan, and sufficient syrup is added so that the finished mass makes a fairly stiff paste. This is then passed through a 12-mesh sieve and evenly distributed in thin layers over shallow trays. These are placed in tiers in a drying cupboard kept at a temperature of about 120°F. When the moisture has evaporated, the trays are removed and the granules again passed through a sieve of smaller mesh, say 16 or 20. From 1 to 5 per cent of talc is now mixed in with the granules; this acts as a lubricant, yielding a smoother flow on the machine. The perfume is evenly distributed and the whole transferred, a few pounds at a time, to the hopper of the machine. For economical production a punch rotary tablet machine is employed, and the punches and dies adjusted to yield a tablet weighing 4 or 5 grains.

CHAPTER ELEVEN

Incense and Fumigants

The burning of aromatic plants and their products has from time immemorial played an important part in religious ceremonies. The monuments of ancient Assyria and Egypt bear ample proof of this by the numerous sculptures and paintings appearing on them. There are further references to this practice in the writings of the Scriptures. For instance, in Exodus xxx 1: 'And thou shalt make an altar to burn incense upon.' Other references are to be found in Exodus xxx 7, 8, 34; 2 Kings xxii 10, 11; xxiii 5. The Egyptians burned the incense in censers and used principally myrrh, frankincense, and a specially blended mixture of sixteen ingredients called *Kaphi*. While this custom has been abandoned by many churches because of the great esteem in which it is held by the heathens, it still finds a wide use by the Roman Catholic and Greek churches. The principal form in which incense is burnt in the East today is as joss-sticks. These are fashioned somewhat like a candle and are said to be made principally of powdered santalwood (both East Indian and West Australian) mixed with swine's dung. The incense burned in the European churches today appear to consist mainly of olibanum: some species contain a little charcoal, benzoin, and storax.

Aromatic substances are nowadays burnt also as fumigants in sickrooms and as perfumes for apartments. Powdered woods and spices are used, but where possible they are replaced by essential oils and charcoal. As is well known, the combustion of aromatic woods gives rise to empyreumatic substances which have a deleterious influence on the odours of the other ingredients.

Charcoal, on the other hand, possesses none of these disadvantages, and when lighted continues to glow—the heat given off volatilising the aromatic essential oils and gums. Saltpetre is often added to assist combustion.

Fumigating pastilles

Amongst the substances employed in the manufacture of fumigating pastilles are the following:

>   Siam benzoin
>   Tolu balsam
>   Peru balsam
>   Labdanum
>   Cascarilla oil
>   Sandalwood oil
>   Cinnamon bark oil
>   Clove oil
>   Cassia oil
>   Patchouli oil
>   Vetivert
>   Grain musk
>   Civet

The pastilles are made by powdering the gums, adding the charcoal and potassium nitrate, spraying on the essential oils and finally massing with a mucilage of either gum acacia or tragacanth. An example follows:

No. 1281

100	Siam benzoin
50	Tolu balsam
700	Charcoal
50	Saltpetre
50	Sandalwood oil
15	Patchouli oil
30	Cascarilla oil
5	Grain musk
q.s.	Mucilage of acacia
1000	

Press the mass into conical moulds.

Perfumed incense

Perfumed incense is frequently sold in powder form and does not contain olibanum, since the smallest percentage is noticeable. In order to obtain new odour notes, perfumers resort to the use of many of the modern synthetics as well as spices such as cardamon and cubebs. The essential oils and gum-resins are added in such proportions that the finished powder is damp and adherent. It can then be easily pressed into a conical shape, placed on an ash-tray and lighted. An example of this type is given:

No. 1282

150	Siam benzoin
300	Sandalwood, in powder
150	Cascarilla, in powder
120	Willow charcoal
30	Saltpetre
40	Cardamons, in powder
35	Cubebs, in powder
10	Myrrh, in powder
5	Grain musk
30	Bergamot oil
10	Patchouli oil
20	Neroli oil
15	Cassia oil
5	Clove oil
30	Peru balsam
10	Orris oleo-resin
30	Iso-butyl cinnamate
10	Methyl ionone
1000	

Perfumed ribbon

Perfumed ribbon is prepared by immersing rolls of very thin wick in saturated solution of saltpetre and subsequently perfuming. The ribbon is then burned in a special lamp or is often supplied in a box made for the purpose. A suitable perfume is made as follows:

No. 1283

30	Musk ambrette
30	Coumarin

No. 1283 (*continued*)

20	Vanillin
50	Rose otto, synthetic
20	Jasmin, synthetic
50	Vetivert oil
300	Tincture of tolu, 10 per cent
500	Tincture of benzoin, 10 per cent
1000	

Perfumed cards are often used for advertising purposes and are better perfumed before printing. For this purpose the card in sheets is placed in a specially constructed chamber and the perfume volatilised at a low temperature from the shallow tray in the bottom. This method yields uniform results and is better than immersion, when the sheets are apt to crinkle or come out in streaks. The perfume chosen is generally the one being advertised when it should be prepared for the purpose in concentrated form. An example of lavender water is given:

No. 1284

50	Lavender oil, English
150	Lavender oil, French
200	Bergamot oil
20	Thyme oil
30	Rosemary oil
20	Oakmoss absolute
30	Patchouli oil
25	Musk ambrette
25	Coumarin
450	Tincture of benzoin, 10 per cent
1000	

Perfumed programmes may be treated like the latter, but are frequently sprayed on the edges while they stand aside in piles. An example of an amber type is given:

No. 1285

200	Tincture of musk, 3 per cent
50	Tincture of civet, 3 per cent
250	Tincture of ambergris, 3 per cent
25	Labdanum clair
15	Oakmoss absolute decolorised

20	Jasmin compound
40	Rose compound
30	Musk ambrette
40	Vanillin
30	Benzyl iso-eugenol
300	Tincture of benzoin, 10 per cent
1000	

CHAPTER TWELVE

Sachels and Solid Perfumes

Sachet perfumes

Sachet powders are most useful for placing amongst clothes and linen without any fear of damage. They are attractively produced by packing in silk or velvet envelopes, and when skilfully made the odour will last for years. The perfumer is somewhat limited in his choice of a dry odourous natural base, since many aromatic raw materials only retain their odour while fresh; notable exceptions amongst the flowers are lavender, clove, and rose; amongst the leaves: patchouli, orange, and lemon; amongst the roots: vetivert and orris; amongst the woods: sandal, cedar, and rosewood; amongst the seeds: ambrette, tonka, anise, and nutmeg; and amongst the barks: cascarilla, cinnamon, and cassia. In the case of the gum-resins all of these are available, and those which cannot be powdered may be rubbed in or sprayed on in alcoholic solutions. The crystalline synthetics could be employed direct, but it is generally better to use them in any added compound. The animal extracts are most successfully employed in the form of concentrated alcoholic extracts and are always used in the best quality sachets. In a few cases grain musk is added direct. Vanilla may be utilised either way.

There are two separate and distinct methods available for the production of sachets; the first by the use of one general powder base to which is added the distinctive floral compound; the second where the separate natural powder bases are chosen according to their odour and subsequently topped by a compound or suitable synthetics, natural isolates, and where price will allow—flower absolutes. Alternatively cheapness can be obtained by diluting the finished powder to price with maize starch.

A sachet base for the former type is made by mixing the natural substances, previously powdered, in suitable proportions. Almost all of them are available in this state and are stocked by the largest wholesale houses. An example of this type is given:

No. 1286

200	Orris
150	Sandalwood
100	Cedarwood
100	Rosewood
70	Patchouli
50	Vetivert
80	Rose petals
100	Lavender
50	Ambrette seed
30	Tonka beans
59	Siam benzoin
10	Clove
1	Grain musk
1000	

These substances are well mixed and passed through a coarse sieve twice. The base is stored until required in air-tight containers. When finishing off this base ready for packing there are two methods available which differ according to the finished odour desired. For instance, if the sachet is a lavender one, then the procedure would be according to this formula:

No. 1287

600	Sachet base
350	Lavender flowers
50	Lavender Compound
1000	

If the sachet, however, is to have an odour which cannot be enhanced by the further additions of natural aromatic powders, such as heliotrope, the procedure would be as follows:

No. 1288

950	Sachet base
50	Heliotrope Compound
1000	

In this particular instance it would of course be possible to select harmonising ingredients, such as vanilla and benzoin, and add them as in the lavender example, but such a procedure would probably spoil the odour of the already matured compound.

Individual sachets can be compounded very much according to the fancy of the perfumer, the natural powdered substances being selected according to their harmonising qualities. A few standard examples are given:

Carnation, no. 1289

350	Orris
200	Sandalwood
120	Rose petals
30	Tonka beans
50	Siam benzoin
200	Cloves
10	Iso-eugenol
5	Amyl salicylate
1	Mace
5	Terpineol
10	Benzyl iso-eugenol
5	Vanillin
10	Heliotropin
3	Musk ketone
1	Bromstyrole
1000	

Chypre, no. 1290

300	Sandalwood
150	Rose petals
50	Cedarwood
100	Lavender flowers
300	Patchouli leaves
30	Vetivert roots
20	Tonka beans
15	Oakmoss absolute, green
5	Ylang-ylang oil, Bourbon
3	Sassafras oil
2	Dimethyl hydroquinone
1	Labdanum clair
10	Castoreum extract, 3 per cent
4	Vanillin
7	Heliotropin
3	Musk ambrette
1000	

Heliotrope, no. 1291

350	Orris
300	Rose
100	Siam benzoin
100	Tonka beans
100	Heliotropin
10	Vanillin
15	Anisic aldehyde
5	Neroli oil, bigarade
5	Musk ketone
10	Civet extract, 3 per cent
5	Almond oil, S.A.P.
1000	

Jasmin, no. 1292

700	Rosewood
250	Orris
30	Benzyl acetate
10	Ylang-ylang oil, Bourbon
2	Amyl cinnamic aldehyde
3	Musk ketone
5	Civet extract, 3 per cent
1000	

Lavender, no. 1293

450	Lavender flowers
200	Sandalwood
100	Orris
100	Tonka beans
100	Patchouli leaves
20	Lavender oil, M.B.
2	Oakmoss absolute, green
10	Vanilla extract, 10 per cent
5	Petitgrain oil
10	Bergamot oil
3	Musk ketone
1000	

Orange Blossom, no. 1294

600	Rosewood
200	Sandalwood
150	Orris
20	Petitgrain oil, French
2	Geranyl formate

Orange Blossom, no. 1294 (*continued*)

5	Benzyl acetate
5	Cananga
5	Terpineol
8	Methyl naphthyl ketone
3	Musk ketone
2	Orange blossom absolute
1000	

Rose, no. 1295

700	Rose petals
50	Patchouli leaves
30	Siam benzoin
20	Clove
150	Orris
20	Rose geranium oil, French
5	Rose otto
15	Phenylethyl alcohol
3	Musk ketone
5	Ambergris extract, 3 per cent
2	Heliotropin
1000	

Trèfle, no. 1296

200	Sandalwood
200	Orris
300	Rose
100	Lavender
30	Patchouli
20	Vetivert
70	Tonka beans
30	Benzoin, Siam
30	Amyl salicylate
1	Oakmoss absolute, green
1	Clary sage, concrete
3	Ylang-ylang oil, Bourbon
5	Jasmin, compound
10	Musk ambrette
1000	

Violet, no. 1297

400	Orris
300	Sandalwood
200	Cedarwood

30	Vetivert
20	Siam benzoin
30	Methyl ionone
1	Violet leaves absolute
10	Heliotropin
3	Jasmin, compound
6	Musk ambrette
1000	

Pot-pourri differs mainly from the above in that the raw materials are mixed whole, i.e. *unground*. No particular type of odour is imitated other than a sweet persistent bouquet. The addition of a flower compound is entirely optional. An example follows:

No. 1298

400	Red rose petals
100	Lavender flowers
200	Coarse sandalwood
50	Coarse orris rhizome
30	Patchouli leaves
10	Vetivert roots
10	Tonka beans, broken
20	Cloves
10	Cinnamon bark
10	Allspice
10	Mace
50	Calamus root
30	Benzoin, Siam
20	Coriander seeds
10	Rose artificial
10	Neroli artificial
10	Jasmin artificial
10	Amber, synthetic
10	Heliotropin
1000	

Olla-Podrida might be termed the 'stock-pot' of the perfumery manufacturer. It consists of all the waste materials and spent plant and animal residues from which alcoholic extracts are prepared. These are made saleable by the addition of cheap herbs such as thyme, rosemary, lavender, and rose petals.

Solid perfumes

It is well known that the best way of presenting perfumes is in the ethyl alcoholic form, but a method which has recently suggested itself is that of using a fat or wax base as an inodorous vehicle. On the face of it No. 36 flower pomade would seem to be the ideal medium, but since it is produced from a mixture of two of lard to one of beef suet, it is naturally a rather greasy base. This can be lessened to some extent by making suitable additions of beeswax when the concentrated flower oil is added to the melted pomade as follows:

No. 1299

600	No. 36 rose pomade
200	White beeswax
200	Concentrated flower oil
1000	

This makes a fairly satisfactory solid perfume but the objection of some greasiness is still present. With a view to the production of a suitable non-greasy product the author carried out a number of experiments and obtained completely satisfactory results by utilising a mixture of Japan wax and white beeswax together with ethyl phthalate and any fancied concentrated floral compound as follows:

No. 1300

200	White beeswax
200	Japan wax
400	Ethyl phthalate
200	Concentrated flower oil
1000	

By varying the proportions of wax and fixative, hard or soft products may be obtained, and the perfume base can, of course, be chosen according to the fancy of the perfumer.

CHAPTER THIRTEEN

Fruit Flavours

Wholly artificial fruit flavours, however skilfully prepared, always lack that finish which characterises those containing a proportion of natural fruit juice. This should be borne in mind when preparing the flavours described below, and doubtless price will influence the proportion which may be subsequently added.

Fruit juices

The preparation of the natural juice involves the crushing of the fruits between stone rollers where metal is not allowed to come into contact with the product. (A reaction would take place between the acids present in the juice and the metal, forming soluble salts which might interfere with both taste and colour.) In some cases, notably those of pineapple and strawberry, the fresh fruit is crushed and the juice pressed out and filtered. In others, notably those of apples, plums, pears, and apricots, the flavour is much improved by fermentation which is effected before the pressure of the pulp. For this purpose from 2 to 5 per cent of sugar is added, and after inversion by the fruit acids, fermentation proceeds at a temperature of 22° to 30°C the necessary time for complete development of the alcohol being from five to seven days. As the alcohol is formed, so is the pectin precipitated. This not only clarifies the juice but also much facilitates its filtration. Media sometimes used for filtration include, kaolin, kieselguhr, talcum, and asbestos. One-half per cent of skimmed milk is also said to be a useful addition. The fruit juice may now be sterilised to destroy bacteria and thus prevent deterioration. This is accomplished by heating the juice in a closed apparatus to about 82°C for half an hour. Incidentally, this coagulates any albumin present and thus further facilitates clarification, which takes place in sealed containers. The use of organic substances, such as sodium benzoate, benzoic and salicylic acids, as preservatives is only allowed in certain countries. The conversion of the juice into a syrup by solution of about 60 per cent of sugar (13 sugar to 7 juice) is another method, not infrequently adopted because most juices are sweetened before being used.

The yield of juices for the fruits is approximately as follows:

	per cent
Apples	65
Bilberries	70
Blackberries	80
Cherries	70
Currants, Black	70
Currants, Red	80
Currants, White	70
Gooseberries	70
Grapes	75
Pears	70
Rasberries	75
Strawberries	80

The concentration of fruit juices is really nothing more nor less than a fractional distillation *in vacuo*—the first runnings (about 10 per cent), containing the most volatile aromatic constituents, being added to the residues in the still. To make sure of the clarity of these concentrated juices, it is customary to add some alcohol to the still residues before the first fraction is added again. This precipitates any pectinous matter which is filtered out. The filtrate is transferred again to the still and the alcohol recovered. The first aromatic fractions are then added when a perfectly bright and clear concentrated product results. To obtain 1 litre of concentrated fruit juice the following approximate quantities must be taken:

litres	juice
6	Apple
20	Apricot
8	Bilberry
10	Blackberry
6	Cherry

10	Cranberry
8	Currant
6	Gooseberry
4	Grape
6	Lemon
5	Mandarin
6	Mulberry
4	Orange
6	Pear
20	Peach
6	Pineapple
10	Plum
10	Strawberry

Artificial flavours

Artificial flavours are based very largely upon synthetics of a very volatile nature, and the lack of sustained taste on the tongue will be readily understood. It is therefore customary to use small quantities of other substances having necessarily no flavour relationship, but which act very much in the manner of a perfumery fixative. The most commonly employed substance is vanillin. Heliotropin and also coumarin have their similar uses. Glycerine is often used as a vehicle, an organic acid sharpens the taste and the diluent is generally alcohol.

Apple components—Acetaldehyde, amyl acetate, butyrate and valerianate, butyl aldehyde, butyrate, formate and valerianate, ethyl acetate, butyrate, malonate, propionate, nitrite and valerianate, ethyl oenanthate, chloroform, geranyl butyrate, isobutyl acetate and valerianate, phenylethyl butyrate and valerianate, phenylglycol formate, vanillin, benzaldehyde, eugenol, clove oil and pimento oil, petitgrain oil, octyl and undecyl acetates.

Apple Base, no. 1301

300	Amyl valerianate
200	Ethyl malonate
100	Acetaldehyde
50	Chloroform
45	Geranyl butyrate
200	Ethyl acetate
5	Vanillin
100	Glycerine
1000	

Apricot components—Acetaldehyde, amyl alcohol, acetate, butyrate and formate, anethol, benzaldehyde, chloroform, benzyl acetate, cinnamate, formate and propionate, ethyl acetate, butyrate, formate, cinnamate and valerianate, methyl cinnamate, ethyl oenanthate, phenylethyl acetate, butyrate and formate, gamma undecalactone, jasmin absolute, methyl salicylate, ethyl salicylate, vanillin, tartaric acid, pimento oil.

Apricot Base, no. 1302

70	Amyl butyrate
150	Benzaldehyde
300	Ethyl butyrate
200	Ethyl valerianate
50	Ethyl salicylate
30	Ethyl oenanthate
50	Chloroform
10	Gamma undecalactone
10	Vanillin
25	Petitgrain oil, French
5	Jasmin absolute
100	Glycerine
1000	

Banana components—Amyl acetate and butyrate, benzaldehyde, benzyl acetate, formate, propionate and cinnamate, cinnamon, clove, coriander, camomile, ethyl acetate, butyrate and sebacate, ethyl oenanthate, lemon, orange, petitgrain, eugenol, pimento, vanillin.

Banana Base, no. 1303

300	Amyl acetate
100	Amyl butyrate
15	Benzaldehyde
40	Benzyl propionate
400	Ethyl butyrate
100	Ethyl sebacate
20	Clove oil
20	Petitgrain oil
5	Vanillin
1000	

Blackberry components—Amyl acetate, ethyl acetate, butyrate, benzoate and oenanthate, methyl salicylate, ionone, orris, vanillin, heliotropin, benzoic acid.

Blackberry Base, no. 1304

400	Ethyl acetate
100	Ethyl butyrate
200	Ethyl benzoate
100	Ethyl oenanthate
60	Methyl ionone
40	Methyl salicylate
50	Orris concrete
10	Vanillin
40	Benzoic acid
1000	

Cherry components. Amyl alcohol, acetate, butyrate and formate, benzaldehyde, ethyl acetate, benzoate, butyrate, oenanthate and pelargonate, cassia, cinnamon, clove, benzoic acid, pimento, vanillin, petitgrain, ethyl cinnamate, glycerine, lemon, orange, gamma undecalactone.

Cherry Base, no. 1305

50	Amyl formate
50	Amyl butyrate
300	Ethyl acetate
400	Ethyl benzoate
100	Ethyl oenanthate
5	Cinnamon bark oil
10	Clove oil
20	Petitgrain oil
5	Gamma undecalactone
10	Vanillin
50	Benzoic acid
1000	

Gooseberry components—Acetaldehyde, amyl acetate and formate, benzoic acid, benzaldehyde, coumarin, ethyl acetate, butyrate, formate, benzoate and oenanthate, geraniol, lemon, clove, petitgrain, pimento, fennel, rose, sassafras, glycerine, succinic and tartaric acids.

Gooseberry Base, no. 1306

50	Acetaldehyde
200	Amyl formate
300	Ethyl acetate

Gooseberry Base, no. 1306 (*continued*)

200	Ethyl butyrate
100	Ethyl benzoate
20	Geraniol
30	Succinic acid
5	Sassafras oil
5	Fennel oil
10	Coumarin
80	Glycerine
1000	

Grape components—Acetaldehyde, amyl acetate and butyrate, cardamon, clary sage, chloroform, cinnamyl propionate, ethyl acetate, anthranilate, formate and oenanthate, diethyl succinate, methyl salicylate, benzoic, succinic and tartaric acids, octyl isobutyrate and valerianate, rhodinol, mace, ethyl pelargonate, terpineol, vanillin, glycerine, cognac (ethyl heptoate), clove, lemon, orange.

Grape Base, no. 1307

50	Acetaldehyde
50	Amyl butyrate
50	Chloroform
10	Clary sage
400	Ethyl acetate
200	Ethyl oenanthate
30	Ethyl pelargonate
50	Ethyl formate
10	Methyl salicylate
30	Succinic acid
10	Cognac oil
5	Octyl valerianate
5	Vanillin
100	Glycerine
1000	

Melon components—Acetaldehyde, amyl acetate, butyrate and valerianate, benzyl acetone, ethyl acetate, butyrate, oenanthate, formate, pelargonate, capronate, sebacate and valerianate, eugenol, cassia, clove, pimento, vanillin, lemon, orange, glycerine, undecalactone.

Melon Base, no. 1308

150	Ethyl acetate
100	Ethyl formate
150	Ethyl butyrate
400	Ethyl valerianate
50	Ethyl pelargonate
4	Benzyl acetone
5	Eugenol
1	Gamma undecalactone
100	Ethyl sebacate
30	Lemon oil
10	Vanillin
1000	

Peach components—Amyl alcohol, acetate, butyrate and formate, benzaldehyde, benzyl cinnamate, acetaldehyde, butyric aldehyde, cinnamyl acetate, ethyl acetate, butyrate, cinnamate, formate, valerianate, oenanthate and sebacate, geranyl butyrate, iso-butyl cinnamate, methyl cinnamate and salicylate, phenylethyl acetate, butyrate and iso-butyrate, undecalactone, vanillin, lemon, orange, clove, pimento, cinnamon, cardamon, petitgrain.

Peach Base, no. 1309

500	Gamma undecalactone
150	Amyl acetate
50	Amyl formate
10	Benzaldehyde
40	Benzyl cinnamate
50	Ethyl oenanthate
50	Ethyl butyrate
50	Ethyl valerianate
100	Vanillin
1000	

Pear components—Amyl acetate, formate and valerianate, ethyl acetate and butyrate, butyl butyrate, iso-butyl acetate, ethyl oenanthate, vanillin, eugenol, bergamot, orange, petitgrain, methyl salicylate, octyl acetate, duodecyl acetate, glycerine.

Pear Base, no. 1310

400	Amyl acetate
400	Ethyl acetate

Pear Base, no. 1310 (continued)

70	Ethyl butyrate
70	Sweet orange oil
30	Bergamot oil
5	Eugenol
20	Vanillin
5	Methyl salicylate
1000	

Pineapple components—Acetaldehyde, amyl acetate, butyrate and formate, benzyl formate and propionate, chloroform, ethyl acetate, butyrate, formate and propionate, ethyl oenanthate, methyl cinnamate, phenylethyl valerianate, ethyl pelargonate and sebacate, vanillin, lemon, orange, petitgrain, pimento, camomile, glycerine, propyl valerianate, ethyl methyl phenylglycidate.

Pineapple Base, no. 1311

500	Amyl butyrate
200	Ethyl butyrate
50	Ethyl acetate
60	Acetaldehyde
50	Chloroform
20	Lemon oil
10	Ethyl methyl phenylglycidate
100	Propyl valerianate
10	Vanillin
1000	

Plum components—Acetaldehyde, amyl acetate and butyrate, benzaldehyde, ethyl acetate, formate, butyrate, oenanthate, methyl formate, undecalactone, gamma nonyl lactone, clove, pimento, coriander, lemon, mandarin, vanillin, glycerine.

Plum Base, no. 1312

250	Acetaldehyde
250	Ethyl acetate
100	Ethyl butyrate
50	Ethyl formate
200	Ethyl oenanthate
20	Benzaldehyde
5	Gamma nonyl lactone
10	Clove oil

10	Mandarin oil
5	Vanillin
100	Glycerine
1000	

Raspberry components—Amyl acetate, butyrate and formate, benzyl acetone, butyl formate, benzyl acetate, clove, cinnamon, ethyl acetate, benzoate, butyrate, caprylate, formate, sebacate, nitrite and oenanthate, iso-butyl acetate and formate, ionone, methyl salicylate, orris, mace, vanillin, heliotropin, benzoic, succinic and tartaric acids, glycerine, gamma nonyl lactone.

Raspberry Base, no. 1313

300	Iso-butyl acetate
200	Amyl acetate
100	Ethyl acetate
10	Clove oil
50	Ethyl butyrate
50	Ethyl formate
10	Ethyl benzoate
50	Ethyl oenanthate
30	Ethyl nitrite
70	Acetaldehyde
10	Methyl salicylate
20	Ionone alpha
50	Succinic acid
30	Gamma nonyl lactone
20	Vanillin
1000	

Strawberry components—Ethyl methyl phenylglycidate, amyl acetate and butyrate, benzyl acetone, ethyl acetate, butyrate, cinnamate, formate, nitrite, anthranilate, salicylate, pelargonate and oenanthate, geranyl butyrate, ionone, methyl salicylate, phenylethyl propionate, orris, geraniol, vanillin, coumarin, methyl naphthyl ketone, cinnamon, succinic acid, glycerine.

Strawberry Base, no. 1314

100	Ethyl methyl phenylglycidate
300	Ethyl acetate
30	Ethyl benzoate
200	Ethyl butyrate

Strawberry Base, no. 1314 (*continued*)

100	Ethyl nitrite
50	Ethyl perlargonate
100	Ethyl formate
40	Amyl acetate
30	Benzyl acetone
10	Methyl naphthyl ketone
20	Methyl salicylate
10	Cinnamon oil
10	Coumarin
1000	

APPENDIX

Conversion Tables

1. Measures of weight—grams into avoirdupois.
2. Grams per kilo into grains per pound.
3. Grams per litre into grains, etc., per fluid ounce, pint, and gallon.
4. Cc. per litre into minims, etc., per fluid ounce, pint and gallon.
5. Measures of capacity—cc. into pints.
6. Conversion data.
7. Fineness of powders.

Measures of weight

1 gram	= the weight of 1 cubic centimetre (cc or mill) of water at $4°C$
1 grain	= 0·0648 gram
1 oz. (Troy)	= 31·1035 grams
1 lb. avoirdupois	= 453·593 grams

	Grams	Grains	Avoirdupois		
			lb	oz	drams
Milligram	0·001	0·0154			
Centigram	0·01	0·1543			
Decigram	0·1	1·5432			
Gram	1·0	15·4323			
Decagram	10·0	154·3234	0	0	5·65
Hectogram	100·0	1543·2348	0	3	8·5
Kilogram	1000·0	15432·3488	2	3	5

Grams per kilo (*parts per 1000*)

Into grains, etc., per lb (*7000 grains*)

Grams per kilo	Per lb	
	oz	grains (avoir)
1	0	7·0
2	0	14·0
3	0	21·0
4	0	28·0
5	0	35·0
6	0	42·0
7	0	49·0
8	0	56·0
9	0	63·0
10	0	70·0
20	0	140·0
30	0	210·0
40	0	280·0
50	0	350·0
60	0	420·0
70	1	52·5
80	1	122·5
90	1	192·5
100	1	262·5
200	3	87·5
300	4	350·0
400	6	175·0
500	8	—
600	9	262·5
700	11	87·5
800	12	350·0
900	14	175·0
1000	16	—

APPENDIX

Grams per litre

Into grains, etc., per fluid ounce, pint, and gallon

Grams per litre	Grains per fluid oz	Grains, etc., per pint		Grains, etc., per gallon			Grams per litre
		oz	grains	lb	oz	grains	
1	0·43	0	8·75	0	0	70·0	1
2	0·87	0	17·50	0	0	140·0	2
3	1·31	0	26·25	0	0	210·0	3
4	1·75	0	35·00	0	0	280·0	4
5	2·18	0	43·75	0	0	350·0	5
6	2·62	0	52·50	0	0	420·0	6
7	3·06	0	61·25	0	1	52·5	7
8	3·50	0	70·00	0	1	122·5	8
9	3·93	0	78·75	0	1	192·5	9
10	4·37	0	87·50	0	1	262·5	10
20	8·75	0	175·00	0	3	87·5	20
30	13·12	0	262·50	0	4	350·0	30
40	17·50	0	350·00	0	6	175·0	40
50	21·87	1	0	0	8	0	50
60	26·25	1	87·50	0	9	262·5	60
70	30·62	1	175·00	0	11	87·5	70
80	35·00	1	262·50	0	12	350·0	80
90	39·37	1	350·00	0	14	175·0	90
100	43·75	2	0	1	0	0	100
200	87·50	4	0	2	0	0	200
300	131·25	6	0	3	0	0	300
400	175·00	8	0	4	0	0	400
500	218·75	10	0	5	0	0	500
600	262·50	12	0	6	0	0	600
700	306·25	14	0	7	0	0	700
800	350·00	16	0	8	0	0	800
900	393·75	18	0	9	0	0	900
1000	437·50	20	0	10	0	0	1000

Cubic centimetres per litre

Into minims, etc., per fluid ounce, pint, and gallon

cc per litre	Per fluid oz	Per pint		Per gallon			cc per litre
	minims	fluid oz	minims	pints	fluid oz	minims	
1	0·48	0	9·6	0	0	76·8	1
2	0·96	0	19·2	0	0	153·6	2
3	1·44	0	28·8	0	0	230·4	3
4	1·92	0	38·4	0	0	307·2	4
5	2·40	0	48·0	0	0	384·0	5
6	2·88	0	57·6	0	0	460·8	6
7	3·36	0	67·2	0	1	57·6	7
8	3·84	0	76·8	0	1	134·4	8
9	4·32	0	86·4	0	1	211·6	9
10	4·80	0	96·0	0	1	288·0	10
20	9·60	0	192·0	0	3	96·0	20
30	14·40	0	288·0	0	4	384·0	30
40	19·20	0	384·0	0	6	192·0	40
50	24·00	1	0	0	8	0	50
60	28·80	1	96·0	0	9	288·0	60
70	33·60	1	192·0	0	11	96·0	70
80	38·40	1	288·0	0	12	384·0	80
90	43·20	1	384·0	0	14	192·0	90
100	48·00	2	0	0	16	0	100
200	96·00	4	0	1	12	0	200
300	144·00	6	0	2	8	0	300
400	192·00	8	0	3	4	0	400
500	240·00	10	0	4	0	0	500
600	288·00	12	0	4	16	0	600
700	336·00	14	0	5	12	0	700
800	384·00	16	0	6	8	0	800
900	432·00	18	0	7	4	0	900
1000	480·00	20	0	8	0	0	1000

APPENDIX

Measures of capacity

1 litre = 1 cubic decimetre = 35·214 fluid ounces
1 gallon = 4·54596 litres

	Litres	Pints
Millilitre (cc. or mil.)	0·001	0·0017
Centilitre	0·01	0·0176
Decilitre	0·1	0·1760
Litre	1·0	1·7607
Decalitre	10·0	17·6077
Hectolitre	100·0	176·0773
Kilolitre	1000·0	1760·7734

Conversion data

Grams × 15·432 = grains
Grains × 0·0648 = grams
Ounces × 28·349 = grams
Pints × 567·936 = cubic centimetres
Gallons × 4·548 = litres
Litres × 0·22 = gallons

$$\frac{\text{Grains per gallon}}{0·7} = \text{parts per 100,000}$$

Parts per 100,000 × 0·7 = grains per gallon

$$\text{Degrees Twaddell} = \frac{1000 \text{ (specific gravity)} - 1000}{5}$$

$$\text{Specific Gravity} = \frac{(\text{Degrees Twaddell} \times 5) \times 1000}{1000}$$

Fineness of powders

Diameter of particles passing through a

No. 40 mesh sieve is less than 0·38 millimetre
No. 50 mesh sieve is less than 0·28 millimetre
No. 60 mesh sieve is less than 0·23 millimetre
No. 80 mesh sieve is less than 0·17 millimetre
No. 100 mesh sieve is less than 0·14 millimetre
No. 120 mesh sieve is less than 0·12 millimetre
No. 150 mesh sieve is less than 0·09 millimetre
No. 200 mesh sieve is less than 0·07 millimetre

TABLE I

Specific gravity of mixtures of alcohol and water

Specific Gravity at 60°F (15.5°C)	Absolute alcohol		Percentage of proof spirit	Specific gravity at 60°F (15.5°C)	Absolute alcohol		Percentage of proof spirit
	By volume	By weight			By volume	By weight	
1000	0.00	0.00	0.00	965	30.34	24.97	53.04
999	0.66	0.53	1.16	964	31.18	25.68	54.51
998	1.34	1.07	2.33	963	31.99	26.37	55.93
997	2.02	1.61	3.52	962	32.79	27.06	57.33
996	2.72	2.17	4.73	961	33.56	27.73	58.68
995	3.42	2.73	5.98	960	34.33	28.39	60.03
994	4.14	3.31	7.24	959	35.06	29.03	61.32
993	4.88	3.90	8.51	958	35.79	29.66	62.60
992	5.63	4.51	9.82	957	36.50	30.28	63.85
991	6.40	5.13	11.16	956	37.20	30.90	65.09
990	7.18	5.76	12.53	955	37.89	31.50	66.29
989	7.98	6.41	13.94	954	38.57	32.09	67.48
988	8.80	7.08	15.38	953	39.22	32.67	68.62
987	9.65	7.76	16.85	952	39.87	33.25	69.76
986	10.51	8.46	18.34	951	40.50	33.81	70.87
985	11.40	9.18	19.87	950	41.13	34.37	71.98
984	12.29	9.91	21.44	949	41.74	34.92	73.05
983	13.20	10.65	23.02	948	42.35	35.46	74.12
982	14.13	11.42	24.66	947	42.95	36.00	75.17
981	15.08	12.20	26.32	946	43.54	36.54	76.21
980	16.04	12.99	27.99	945	44.13	37.07	77.24
979	17.02	13.80	29.70	944	44.71	37.60	78.26
978	18.00	14.61	31.42	943	45.28	38.12	79.26
977	18.99	15.43	33.15	942	45.85	38.64	80.26
976	19.98	16.25	34.87	941	46.40	39.15	81.23
975	20.97	17.08	36.61	940	46.95	39.65	82.19
974	21.96	17.90	38.35	939	47.50	40.15	83.15
973	22.94	18.72	40.06	938	48.04	40.65	84.10
972	23.91	19.53	41.77	937	48.57	41.15	85.04
971	24.85	20.34	43.47	936	49.10	41.64	85.97
970	25.83	21.14	45.14	935	49.63	42.13	86.89
969	26.77	21.93	46.77	934	50.15	42.62	87.81
968	27.69	22.71	48.38	933	50.67	43.11	88.71
967	28.69	23.48	49.98	932	51.18	43.59	89.61
966	29.48	24.23	51.53	931	51.68	44.06	90.49

TABLE I—*continued*

Specific gravity of mixtures of alcohol and water

Specific gravity at 60°F (15·5°C)	Absolute alcohol		Percentage of proof spirit	Specific gravity at 60°F (15·5°C)	Absolute alcohol		Percentage of proof spirit
	By volume	By weight			By volume	By weight	
930	52·18	44·53	91·36	897	67·08	59·37	117·54
929	52·67	45·00	92·93	896	67·50	59·80	118·26
928	53·16	45·47	93·09	895	67·92	60·23	118·98
927	53·65	45·94	93·95	894	68·33	60·66	119·70
926	54·14	46·40	94·80	893	68·74	61·09	120·42
925	54·62	46·87	95·65	892	69·14	61·52	121·14
924	55·10	47·33	96·49	891	69·55	61·95	121·85
923	55·58	47·79	97·33	890	69·95	62·38	122·56
922	56·05	48·25	98·16	889	70·35	62·81	123·27
921	56·52	48·71	98·98	888	70·75	63·24	123·97
920	56·99	49·17	99·80	887	71·15	63·67	124·06
917·76	57·10	49·28	100·00	886	71·55	64·10	125·37
				885	71·95	64·53	126·07
919	57·46	49·63	100·62	884	72·34	64·96	126·77
918	57·92	50·08	101·43	883	72·24	65·39	127·46
917	58·38	50·53	102·24	882	73·13	65·81	128·14
916	58·83	50·98	103·05	881	73·52	66·24	128·82
915	59·29	51·43	103·84	880	73·91	66·66	129·50
914	59·74	51·88	104·63	879	74·30	67·09	130·18
913	60·19	52·33	105·42	878	74·68	67·51	130·86
912	60·63	52·77	106·20	877	75·06	67·93	131·53
911	61·07	53·21	106·97	876	75·44	68·35	132·19
910	61·51	53·65	107·74	875	75·82	68·77	132·86
909	61·95	54·10	108·52	874	76·19	69·19	133·53
908	62·39	54·54	109·29	873	76·57	69·62	134·19
907	62·83	54·98	110·06	872	76·94	70·04	134·84
906	63·26	55·42	110·82	871	77·32	70·46	135·50
905	63·70	55·87	111·59	870	77·69	70·88	136·16
904	64·13	56·31	112·35	869	78·06	71·30	136·81
903	64·56	56·75	113·10	868	78·43	71·72	137·46
902	64·98	57·18	113·84	867	78·80	72·14	138·10
901	65·41	57·62	114·59	866	79·17	72·55	138·74
900	65·83	58·06	115·33	865	79·53	72·97	139·38
899	66·25	58·50	116·07	864	79·89	73·39	140·02
898	66·67	58·93	116·81	863	80·25	73·81	140·65

TABLE I—continued

Specific gravity of mixtures of alcohol and water

Specific gravity at 60°F (15·5°C)	Absolute alcohol		Percentage of proof spirit	Specific gravity at 60°F (15·5°C)	Absolute alcohol		Percentage of proof spirit
	By volume	By weight			By volume	By weight	
862	80·61	74·22	141·28	827	91·98	88·27	161·26
861	80·97	74·64	141·91	826	92·26	88·65	161·76
860	81·32	75·05	142·54	825	92·55	89·03	162·26
859	81·68	75·47	143·16	824	92·83	89·41	162·75
858	82·03	75·88	143·78	823	93·11	89·79	162·24
857	83·38	76·30	144·40	822	93·38	90·16	163·72
856	83·73	76·71	145·01	821	93·65	90·53	164·20
855	83·08	77·12	145·62	820	93·92	90·90	164·67
854	83·42	77·53	146·23	819	94·19	91·27	165·14
853	83·77	77·94	146·83	818	94·45	91·63	165·60
852	84·11	78·35	147·43	817	94·71	92·00	166·06
851	84·44	78·76	148·03	816	97·97	92·36	166·51
850	84·78	79·17	148·62	815	95·22	92·72	166·96
849	85·12	79·58	149·21	814	95·47	93·08	167·41
848	85·46	79·98	149·80	813	95·72	93·44	167·86
847	85·80	80·39	150·39	812	95·97	93·80	168·28
846	86·12	80·79	150·97	811	96·21	94·15	168·71
845	86·44	81·20	151·55	810	96·45	94·50	169·13
844	86·77	81·60	152·12	809	96·69	94·85	169·55
843	87·09	82·00	152·68	808	96·93	95·20	169·96
842	87·42	82·40	153·25	807	97·16	95·55	170·37
841	87·74	82·80	153·81	806	97·39	95·89	170·77
840	88·06	83·20	154·37	805	97·62	96·23	171·17
839	88·37	83·60	154·92	804	97·84	96·57	171·56
838	88·68	83·99	155·47	803	98·06	96·91	171·95
837	88·99	84·39	156·02	802	98·28	97·25	172·23
836	89·30	84·78	156·56	801	98·49	97·59	172·71
835	89·61	85·17	157·10	800	98·70	97·91	173·07
834	89·91	85·56	157·63	799	98·91	98·24	173·44
833	90·22	85·95	158·16	798	99·12	98·57	173·80
832	90·52	86·34	158·68	797	99·32	98·90	174·16
831	90·82	86·73	159·21	796	99·52	99·22	174·52
830	91·11	87·11	159·73	795	99·72	99·55	174·87
829	91·40	87·50	160·24	794	99·92	99·87	175·21
828	91·69	87·88	160·75	793·59	100·00	100·00	175·35

TABLE II

Dilution of alcohol by volume with distilled water

Percentage strength of alcohol required by volume	Add to 1000 of alcohol at								
	90	85	80	75	70	65	60	55	50
	Per cent by volume								
85	66								
80	138	69							
75	219	145	72						
70	311	231	153	77					
65	414	330	247	164	82				
60	537	445	354	265	176	88			
55	679	579	481	383	286	190	95		
50	847	739	630	524	417	313	205	104	
45	1053	933	814	695	578	461	345	205	114
40	1308	1173	1040	908	776	645	514	385	256
35	1633	1480	1329	1178	1029	880	700	583	436
30	2062	1886	1711	1535	1363	1189	1017	845	675
25	2661	2452	2243	2036	1828	1622	1417	1212	1007
20	3558	3298	3040	2783	2526	2270	2014	1760	1506
15	5053	4710	4369	4028	3689	3349	3011	2673	2336
10	8045	7537	7029	6522	6016	5511	5005	4502	2999

Examples: to convert 90 per cent to 45 per cent add to 1 litre 1053 cc. Aqua Destil.
to convert 75 per cent to 20 per cent add to 1 litre 2783 cc. Aqua Destil.

TABLE III

Dilution of alcohol by weight with distilled water

Percentage strength of alcohol used	To produce 1000 of alcohol at				
	50	60	70	80	90
	Per cent by weight				
96	453	555	665	783	913
95	460	564	676	796	927
94	467	573	686	808	942
93	474	582	697	820	956
92	481	590	707	832	970
91	489	599	718	845	985
90	496	609	728	858	
89	504	618	740	871	
88	511	627	752	884	
87	519	637	763	898	
86	527	646	774	912	
85	535	656	786	926	
84	543	667	798	940	
83	552	677	811	955	
82	560	687	823	969	
81	569	698	836	984	
80	578	709	849		
79	587	720	863		
78	597	732	877		
77	606	744	891		
76	616	756	905		
75	626	768	920		
74	636	781	935		
73	647	794	951		
72	658	807	967		
71	669	821	983		
70	681	835			
69	692	849			
68	705	864			
67	717	880			

TABLE III—*continued*

Dilution of alcohol by weight with distilled water

Percentage strength of alcohol used	To produce 1000 of alcohol at				
	50	60	70	80	90
	Per cent by weight				
66	730	896			
65	743	911			
64	756	928			
63	770	946			
62	785	963			
61	800	981			
60	815				
59	831				
58	847				
57	864				
56	881				
55	901				
54	918				
53	938				
52	958				

Examples: To make a kilo of 90 per cent by weight take
(1) 913 grams of 96 per cent alcohol and add distilled water to make 1000 grams, or
(2) 956 grams of 93 per cent.

Alcohol table (to required weight and volume)

Showing the strength of alcohol (percentage absolute) with its specific gravity, also the weight or volume required to make one kilogram or one litre of another strength by the addition of distilled water at 15·5° centigrade or 60° fahrenheit

Strength of absolute alcohol	Specific gravity by		Quantity of alcohol of known strength required to make 1000									
			50 per cent by		60 per cent by		70 per cent by		80 per cent by		90 per cent by	
	Weight	Volume	Weight grams	Volume cc	Weight grams	Volume cc	Weight grams	Volume cc	Weight grams	Volume cc	Weight grams	Volume cc
per cent												
96	0·8065	0·8125	453	520	555	625	665	730	783	833	913	935
93	0·8149	0·8237	474	537	582	645	697	752	820	860	956	967
90	0·8229	0·8339	496	555	609	666	728	777	858	888		
85	0·8359	0·8496	535	588	656	705	786	823	926	941		
80	0·8484	0·8639	578	602	709	750	849	875				
75	0·8605	0·8773	626	666	768	800	920	933				
70	0·8724	0·8900	681	701	835	852						
65	0·8842	0·9021	743	769	911	923						
60	0·8958	0·9134	815	833								
55	0·9072	0·9242	901	981								

Index

Abram, 5
Abronia, 217
Absolutes, 36
Acacia, 68
   Cavenia, 77
   dealbata, 68
   Farnesiana, 77
   floribunda, 136
   (Mimosa) species, 136
   soap, 283
Act of 1770, 12
Adulterants of rose otto, 191
À la mode, 229
Alcohol, tables, 368
Alcohols, evolution of, 18
Aldehydes, evolution of, 18
Alkalis used in soaps, 269
Almond soap, 284
Amber Cologne, 255
   lavender, 264
   soap, 285
   Synthetic, 217
Ambre Royale aux fleurs, 218
Ambrosia, 218
Amic, L., 92
Anatolian rose otto, 181
Antimony, 5
   sulphide, 5
Antiseptic soap, 318
   value of essential oils, 319
Aphrodisiacs, 158
Appendix, 361
Apple flavour, 353
Apricot flavour, 354
Apuleius, 167
Aqua Mellis, 247
Artificial fruit flavours, 353
Attar of roses, 167

Babylon, 7
Balanos, 8
Balm, 6
   of Gilead, 6
Banana flavour, 354
Bdellium, 5

Belladonna lily, 128
Benzoinette, 219
Bertram, 163
Bertrand Frères, 105
Bible cosmetics, 6
   perfumes, 6
Blackberry flavour, 355
Boiling soap, 269
   on strength, 269
Boronia, 219
Bouquet à la Maréchale, 232
   d'Esterhazy, 229
   des Alpes, 219
   des Fleurs, 229
   for cachous, 334
   soap, 287
Bouvardia, 219
Bracken, 89
Broders, J., 242
Brown Windsor soap, 287
Buckingham flowers, 229
Bulgarian rose industry, 171
Burger, A. M., 242
Butaflors, 37
Butane extraction, 37
Buttermilk soap, 286
Buying soap chips, 272
   absolutes, 41

Cachous, 332
Camel dung, 322
Cananga, 220
Carnation, 72
   cachous, 335
   sachets, 346
   soap, 289
Cassie, 79
Cavendish tobacco, 329
Cedar soap, 288
Cerbelaud, R., 235
Cerighelli, 117
Chapman, 2
Charabot, 18, 151
Chassis, 32, 34
   absolutes, 32

Chauvet, 181
Chemist and Druggist, 259
Cherry flavour, 355
Chèvrefeuille, 102
Chiris, 186
Chlorophyll, 19
    functions, 19
    of, 19
Chypre, 81
    compounds of Continental origin, 235 et seq.
    imperial, 236
    sachets, 346
    soap, 290
Cigar flavour, 326
Cigars, 326
    boxes, 326
    perfume for, 327
Cigarettes, 327
    Egyptian flavour, 328
    Turkish flavour, 328
    Virginia flavour, 327
Citronellol in rose oil, 185
Classified odours, 46
Cleopatra, 5, 167
Clover, 200
Cohobation, 176
Cola, F., 244
Cold process soaps, 270
Cologne compounds, 253
    for men, 253
    soap, 292
Colourless absolutes, 35
    jasmin, 119
Concretes, 34
Consumption of flowers, 38
Continental practice, 234
Convallaria, 128
    majalis, 128
Conversion data, 361
Coppens, 108
Coronilla, 220
Corylopsis, 220
Cosmetics, uses of, 13
Crocker, 50
Crops, 21
Cucumber soap, 295
Cultivation of plants, 20
    tobacco, 322
Cuniasse, L., 238
Curd soap, 291
Curing of tobacco, 323
Cyclamen, 84
    des Alpes, 85
    soap, 293

Dades Valley, 180

Daffodil, 141
Damascenone, 194
Damask rose, 169
Decumaria, 221
Defleurage, 32
Dejeans, 81
Dhumez, 122
Dianthus species, 73
Dillenia, 221
Distillation, 22
Dry soap chips, 270
Duration of evaporation, 63
Dyestuffs for soaps, 274

Eau de Berlin, 230
    Brouts, 154
    Cananga, 266
    Colognes, 248
    Fleur d'oranger, 154
    Portugal, 267
Ecuelle process, 26
Egyptian cigarettes, 328
    cosmetics, 5
        analysis of, 5
    perfumes, 3
Elze, 69, 117, 205
Embalming, 4
Enfleurage, 29, 30
    absolutes, 30
Enzyme action, 19
Erica, 221
Essence bouquet, 230
Essential oils, antiseptic value of, 319
Evaporation, 63
Evolution of oils, 18
Exodus, 6
Expression, 25
Extraction, 28, 36
Extraits aux fleurs, 32

Face lifting, 15
    massage, 15
Fagonia, 221
Fancy perfumes, 217
Farina, 248
Farnesian gardens, 77
Fats for enfleurage, 31
    soap-making, 269
Feminis, 248
Fern, 88
    soap, 295
Filmarone, 90
Fineness of powders, 365
Firmenich, 118, 194, 211
Fixation, 461

INDEX

Flake tobacco, 329
Flavio Orsini, 150
Floral cachous, 332
Florida water, 265
Flower consumption, 38
Foin coupé, 148
Forced jonquille, 141
    lilac, 121
    lilies, 128
    narcissus, 141
Formation of constituents of essential oils, 18
    essential oils, 17
Fougère, 91
    soap, 294
Fouquet, H., 240
Frangipanni, 11, 230
French rose industry, 182
"Frozen" Eau de Cologne, 261
Fruit flavours, 351
    juices, 351
        concentrated, 352
Fumigants, 339
Fumigating pastilles, 340

Gamut of odours, 49
Garden, Dr., 92
Gardenia, 92
Garden of delight, 7
    Eden, 5
Garnier, 93, 186
Gattefossé, R. M., 128, 236
Gerhardt, O., 243
Geronimo Rossi, 168
Gildemeister, 186
Gillyflower, 72
Giroflée, 214
Glichitch, 186
Gloves, perfumed, 12, 150
Glucosidal decomposition, 20
Glycerine and cucumber soap, 295
Glycine, 222
Gooseberry flavour, 355
Grape flavour, 356
    hyacinth, 107
Grasse, 13
Greek cosmetics, 10
    perfumes, 8
Guenther, 188
Gulapana, 173
Gulbirlik, 182

Haarmann, 79, 100
Hackforth-Jones, 181
Hancornia, 222
Havana cigars, 326

Hawthorn, 96
    cachous, 335
Hay, 145
Hayfields, 147
Heart's tongue, 88
Heckel, 146
Heine, 117
Helen of Troy, 8
Heliotrope, 99
    cachous, 335
    sachets, 347
    soap, 296
Henderson, 50
Henna, 5
Herb soap, 297
Herodotus, 4
Hesse, 116, 205
Hoejenbos, 108
Hoffmann, 186
Homer, 166
Honey soap, 298
    water, 246
Honeysuckle, 102
Horace, 208
Horse-guards' bouquet, 231
Hortus Kewensis, 169
Hovenia, 231
Howard Carter, 2
Hugonia, 222
Hugues, 34
Huile antique, 32
    Française, 32
Hungary water, 248
Hyacinth, 106
    soap, 299

Idealia, 223
Igolen, G., 133, 181
Iliad, 8
Incense, 339
Ismene, 223
Iso-Jasmone, 118
Italian essence industry, 25
Italian jasmin, 116

Jacinthe, 106
Japanese bouquet, 231
Jasmin, 111
    cachous, 336
    colourless, 119
    plants, 112
    sachets, 347
    soap, 300
    species, 112
Jean Antoine Farina, 248
Jeancard, 151

Jezebel, 6
Jockey club, 231
Jonesia, 223
Jonquille, 142
Joss sticks, 339

Kabushi oil, 132
Kazanlik valley, 168
Kenneth Graham, 5
Kiss Me Quick, 231
Kleinhovia, 224
Kohl, 2
Kölnisches Wasser, 248
Koran, 7, 208
Kummert, 214
K.P.K., 155

Laelia, 224
Laloue, 151
Lanolin, in soaps, 271
Latakia, 322
Lautier, 34
Lavender sachets, 347
    soap, 301
    water, 261
Lazennec, I., 237
Leap year bouquet, 232
Lilac, 120
    absolute, 122
    compounds of Continental origin, 235 et seq.
    soap, 303
Lilium species, 127
Lily, 125
    of the Valley, 128
    soap, 304
Lime blossom, 224
Linnaeus, 158
Liquid CO_2 extraction, 36
Liquorice in tobacco, 325
Lonicera, 102
Lozenge-made cachous, 333
Lucius Apuleius, 167
Luxonne, 209

Maceration, 29, 32
Madonna lily, 127
Magnol, 131
Magnolia, 131
Mahomet, 7, 208
Maidenhair fern, 89
Male fern, 89
Manila cigars, 326
Mann, H., 235
Mary Queen of Scots, 12
Matching a soap perfume, 281

Maubert, 122
May blossom, 96
    soap, 305
Measures of capacity, 365
    weight, 361
Medicated soap, 318
Magaleion, 9
Melilot, 146
Melon flavour, 357
Men's Cologne, 253
Metabolism, 18
Mignonette, 165
Millefleur bouquet, 232
Milling process, 273
Mimosa, 135
    Cologne, 257
"Mistress of the Night", 204
Monimia, 225
Moths, 104
Mousseline, 233
Mouth pellets, 332
Muguet, 131
Müller, 116
Musk, 7
    compound, 233
    soap, 306

Napoleon, 208
Narcissus, 139
    soap, 307
Naves, 186
Nemesia, 225
Nero, 10
Neroli oil, 151
    artificial, 157
New-mown hay, 145
Nicotiana species, 322
Nicotine, 324
Night-scented Stock, 225
Nineveh, 7
Non-alcoholic concentrates, 217

Odontoglossum, 159
Odour classification, 46
Odourless fixatives, 58
Oeillet, 72
Olfactic fatigue, 54
Olla-podrida, 349
Opoponax, 226
    soap, 308
Orange blossom, 148
    sachets, 347
    flower absolute, 151
        water, 153
    flowers, 150
    oil, 153
    pekoe, 150

Orchids, 158
Orchis, 158
Orris oil, condensation, 24
Otto of rose, 167
Ovid, 99

Palm and olive oil soap, 310
Parmantheme, 211
Parmone, 211
Parone, E., 93
Parry, 186
Passiflora, 226
Paul de Feminis, 248
Pavetta, 227
Peach flavour, 357
Pear flavour, 357
Perfume formation, 16
    in plant, 17
Perfumed cards, 342
    gloves, 12, 150
    incense, 341
    programmes, 342
    ribbon, 341
Perfumes for soaps, 275
Persian lilac, 121
Petitgrain oil, 156
Pfeiffer, 118
Physiological influence of perfumes, 18
Pierre Magnol, 131
Piesse, 48
Pineapple flavour, 358
Pine bouquet soap, 310
Plant metabolism, 19
    waxes, 34
Plastic surgery, 15
Plenderleith, 2
Pliny, 11, 102, 158, 167
Plum flavour, 358
Plumier, 131
Polyanthus compound, 233
Pomades, 30
Pomet, 156
Pompons, 78
Pot-pourri, 349
Powder perfume base, 45
Power, 201
Prevarka, 176
Purchase of absolutes, 41

Queen Elizabeth, 11, 72
    Shubad, 5

Rabak, 132
Randia, 227

Raspberry flavour, 359
Reimann, 324
Reiner, 79
Reseda, 162
    geraniol, 164
Resin, in soap, 269
Rimmel, 48
Robertet, 37, 122, 129
Robin Hood, 96
Robinia, 69
Robiquet, 33
Rogerson, 201
Roll tobacco, 330
Roman cassie, 77
    cosmetics, 10
    perfumes, 10
Rondeletia, 233
Rose, 166
    absolute, 171
    analysis, 186
    Anatolian, 181
    blanch, 196
    cachous, 336
    centifolia, 169
    chemistry, 185
    cultivation, 172
    de Bulgarie, 169
        Mai, 169
    direct oil, 176
    distillation, 177
    evaluation, 188
    French, 182
Rose, Moroccan, 180
    of Sharon, 141
    olfactic examination, 191
    otto, discovery of, 167
    sachets, 348
    soap, 311
    synthetic, 190
    water, 176
        oil, 171
    yield, 178
Rosenthal, 188–211
Roure-Bertrand Fils, 92, 151, 184
Ruzicka, 118, 211

Sabetay, 117, 210
Sachets, 344
Salato, 26
Salway, 201
Santal soap, 312
Santolina, 227
Satie, 151
Sawer, 132
Schimmel, 163–79
Schleinger, H., 105

Scorzetta process, 26
Seager, 186
Separation of plant perfumes, 22
Sesame oil, 4
Shag tobacco, 329
Shakespeare, 88
Shamrock, 200
Shaving soap, 270
    perfume, 313
Shenstone, 72
Shittim wood, 6
Snuff, 330
Soap, 268
    antiseptic, 318
    boiling, 269
    buying, 272
    chips, 270
    cold process, 270
    colours, 274
    compounds, 282
    cracking, 274
    dyes, 274
    essential oils in, 276
    flower oils in, 280
    medicated, 318
    milling, 273
    perfumery, 268
    perfumes, 275
    pigments, 274
    plodding, 274
    raw materials, 269
    shaving, 270
    stock, 269
    super-fatted, 273
    synthetics in, 276
    transparent, 271
Soden, 143–79
Solid Eau de Cologne, 261
    perfumes, 344
Song of Solomon, 125
Spanish jasmin, 113
Sponge process, 25
Spugna process, 26
Statistics, 38
Stephanotis, 227
Stibium, 8
Strawberry flavour, 359
Strengths of concentrated waters, 154
Sub rosa, 167
Suetonius, 10
Super-fatted soaps, 273
Sweet alyssum, 145
    pea, 197
        soap, 314
Symbol of marriage, 96
Syringa compound, 228

Tablet-made cachous, 337
Terpeneless oils, 258
Testicle, 158
Testing absolutes, 41
Theophrastus, 8, 9, 72, 84, 126, 139, 166, 207
Tiemann, 100
Tinnea, 228
Tobacco, 321
    allowed constituents, 325
    constituents, 324
    cultivation, 322
    curing, 323
    English, 322
    factories, 325
    flavours, 327
    manufacture, 324
    plants, 321
    prohibited constituents, 326
    varieties, 322
Toilet waters, 246
Tombarel, 155
Tonka beans in tobacco, 325
Tournaire, 135
Trabaud, 117, 210
Transparent soaps, 271
    perfume, 315
Treff, 117
Trèfle, 200
    Cologne, 257
    sachets, 348
    soap, 316
Trovar, 203
Tuberose, 203
Tulip, 234
Tunisian neroli, 157
Turkish cigarettes, 328
    Rose otto, 181
    tobacco, 322
Turnsole, 100
Tuscan jasmin, 112
Tutankhamen, 2

Unguents, 10
UOP fragrances, 133

Vanilla, 160
Verbena soap, 316
Verley, 205
Vernal grass, 145
Villain, 272
Violet, 207
    cachous, 336
    compounds, 213
    ketones, 212

Violet (*contd.*)
   leaves, 216
   matching in soap, 281
   sachets, 348
   soap, 281, 317
   tree, 209
Virginia cigarettes, 327
   tobacco, 322
Volatile solvents, 29, 33
Voluptuous intoxication, 203

Wagner, A., 240
Walbaum, 163, 188
Wallflower, 214
Wattle blossom, 69
Well-known recipes, 228

Werner, 117
White lilac, 125
Winter, F., 239
Winter heliotrope, 100
   lilac, 121
Woodbine, 103
Woodruff, 145

Yacht club, 234
Yardley, 181, 185
Yields of absolutes, 35
   concretes, 35
Yulan, 132

Zaccharewitz, 23
Zoroaster, 7